Alan M. Jacobs, PhD
Youngstown State University

DuWayne O. Porter, MPH, RS
Health Commissioner, Portage County Combined General Health District, Ohio
Northeast Ohio Medical University

ENVIRONMENTAL SCIENCE

Sustainability for the 21st Century

Kendall Hunt
publishing company

Cover image © Shutterstock, Inc.

Kendall Hunt
publishing company

www.kendallhunt.com
Send all inquiries to:
4050 Westmark Drive
Dubuque, IA 52004-1840

ISBN 978-1-4652-5442-9

Printed in the United States of America

Rationale

Environmental Science—Sustainability for the 21st Century is a basic environmental science textbook emphasizing sustainable development. The authors have both academic and professional practice experience, including undergraduate and graduate teaching, public service through government agencies, and consulting for environmental assessment and remediation companies.

Alan M. Jacobs, PhD, PG (Professional Geologist)
Professor
Geological and Environmental Sciences
Youngstown State University, Youngstown, Ohio

DuWayne Porter, MPH, RS (Registered Sanitarian)
Health Commissioner
Portage County (Ohio) Health Department,
Ravenna, Ohio

Contents

Acknowledgments

Two institutions and one special graduate program provided us with the opportunity to develop a textbook suitable for a general audience of students needing a basic course in environmental science, with emphasis on sustainable development. Alan Jacobs' academic home is in the Department of Geological and Environmental Sciences at Youngstown State University (YSU) in Youngstown, Ohio. DuWayne Porter is Health Commissioner of Portage County, Ohio. Both Jacobs and Porter, since 2001, have jointly participated in a Master's of Public Health (MPH) program through the Consortium of Eastern Ohio MPH (CEOMPH). We owe a debt of gratitude to colleagues, students, and the general public whom we serve for their support.

Alan M. Jacobs, PhD
Professor, Youngstown State University, Youngstown, Ohio
Course Director, Environmental Health Sciences, CEOMPH, Rootstown, Ohio

DuWayne O. Porter, MS, RS,
Health Commissioner,
Portage County Health Department, Ravenna, Ohio
Course Co-Director, Environmental Health Sciences, CEOMPH, Rootstown, Ohio

UNIT 1 Introduction to Science and the Environment

1 Environmental Science

Environmental science is the scientific study of our surroundings. It is multidisciplinary, encompassing the disciplines of the physical and life sciences (pure and applied) and even some aspects of the social sciences. Consequently, this includes physics, chemistry, geology, biology, astronomy, soil science, the health sciences (e.g., medicine and toxicology), political science, and economics. When science is applied to the design of solutions to environmental problems, we also incorporate the study of engineering. Associated with environmental science is another study called *environmental studies*, which emphasizes the social sciences, ethics, lifestyle, and environmental policy.

Dynamic Earth

The layman thinks of the Earth, our home, as being stable and predictable. We expect daylight and darkness each day (unless you are in polar regions). We plan on the beginnings and ends of winter, spring, summer, and autumn to fall on specific calendar days and have

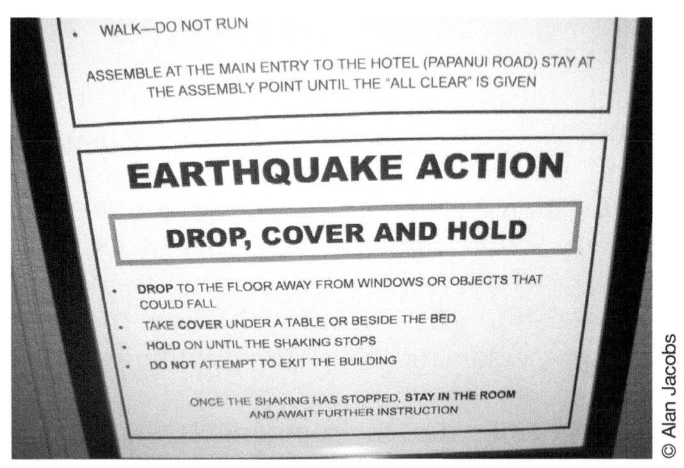

FIGURE 1-1 Hotel earthquake warning, Christchurch, New Zealand

seasonal weather during those seasons. We expect certain amounts of rain, snow, and fair weather from year to year at our location. We expect our environment to be predictable in time and place. If we want changes, we travel or move to other areas of the globe. Once there, we expect a new series of predictable environments.

When we hear of or experience deviations from the norm, however, we categorize such differences as unseasonal, unusual, freakish occurrences, disasters, or even acts of God. We are surprised when there are deviations from the expected norms, and try to blame disparities on something humans are doing wrong—to bring it upon ourselves. We are shocked when our land on floodplains is flooded or when winter snowstorms dump feet of snow in the north or layers of frozen rain in winter in the south. It is newsworthy when hurricanes and tsunamis invade coastal areas, when tornadoes devastate areas in the Midwest, when earthquakes **(Figure 1-1)** and mudslides ravage property in California, and when fires burn acres of droughty grasslands and forests. Are these unexpected occurrences caused by global warming, sunspot activity, phases of the Moon, astrological positioning of planets and star constellations, or our own actions?

We do bring into our environment some unfavorable conditions like pollution, poor management of waste and energy resources, overconsumption, and overpopulation. However, variations in our environment and natural disasters are mostly the result of a dynamic Earth, whose creation has, after 4.6 billion years of existence, not been completed. Will it ever be completed? The Earth will remain a dynamic planet for many billion years to come. Change is the norm.

Land above sea level has been eroded and deposited as sediment in the oceans. Deposition of sediment has been hardened into rock and uplifted into mountain ranges. The rocks of the Earth's crust have been recycled by melting, erosion, deposition, solidification, metamorphism, burial, and exhumation. Plants added oxygen to the atmosphere and oceans with energy from the Sun. Life on Earth, made out of the same elements that have existed on Earth, continues to be maintained by recycling these elements. Populations of organisms changed and became extinct, leaving traces of their existence in the rocks that we find today.[1]

SUSTAINABILITY

A Chinese saying goes like this: Give a person a fish and he or she can eat for a day; teach a person how to fish and he or she can eat for life. Most people attribute this saying to champion education and self-reliance. The current authors also apply this saying to the environment, with modification. Fishing, in this sense, is a metaphor for knowing how to sustain resources necessary for the future. We call this attribute of society "sustainability." What this saying ignores, however, is the problem of future depletion of fish and the possibility of having no more fish to give someone else and, thereby, no need to know how to fish. It also relates to one person and not to the multitudes that require fish, other food, and all material and energy needs.

Sustainability, as it applies to the environment, is the ability to sustain life and its necessities in an ethical and just manner in the near and distant future. The necessities include materials and energy. Materials include food, water, air, medicines, buildings, vehicles for transportation, tools, etc. Energy is needed to move materials, build structures, manufacture products, extract raw materials, and to maintain a safe, healthy, and productive environment.

So the current authors propose a new saying: Let people help themselves to material and energy and they can survive for a while; teach them *environmental sustainability* and they can survive for eons.[2]

In light of these changes, how do we sustain our human species? World populations are growing at alarming rates **(Figure 1-2)**. Our world population of over seven billion is doubling about every three decades at current growth rates. Our technology is just barely keeping up to sustain these numbers. We must accept change and adapt to those changes. Concurrent to this adaptation to existing environmental conditions, we must manage our resources, our wastes, and our lifestyles lest we will join the extinct species that we find now only in the fossil record. The answer is found in the study of environmental science.

Nature has sustained life on Earth for several billion years. Humans have been around for only the last couple million years, and our future existence or dominance on Earth is not guaranteed. The Earth is dynamic, and humans or any other species who do not adapt to changing conditions will not survive.

Nature plays no favorites. Species become extinct. New species arise. The fossil record shows major periods of extinction and periods where different life forms have become dominant.

FIGURE 1-2 Time is running out

FIGURE 1-3 Appearance of primitive humans

About 2.5 billion years ago an oxygen *revolution* occurred,[3] allowing enough oxygen to accumulate to support life. Plants, mostly in the ocean, provided this surplus of oxygen to allow this to happen. They still do.

About 500 million years ago, a change occurred requiring animals (all of them sea creatures) to develop hard parts, exoskeletons (shells).[3] We assume the shells were to protect themselves from predators. The shells were of varied shapes and ornamentations. Some creatures with hard parts were segmented (trilobites). Others had a spiral form or had two-part hinged shells (mollusks and brachiopods). How did previous life forms without shells protect themselves? Were there other reasons for the shells?

About 200 million years ago, life on land started to flourish.[4] What was the impetus? Why did some organisms abandon the sea? Among the most interesting land creatures were the dinosaurs. For tens of millions of years they dominated the Earth. Then 65 million years ago, dinosaurs became extinct. After the demise of the dinosaurs, mammals and birds flourished on land. Several of these, like whales, porpoises, and penguins, returned to the sea (but still breathed air from the atmosphere).

Only a couple million years ago, humans appeared **(Figure 1-3)**.[5] With our superior brain power, we developed the ability to modify our environments and dominate the Earth. Yet, humans are part of the biosphere. We are bound by the same natural laws that control all other life forms. We are so abundant and powerful, however, that we have threatened the balance of Nature in our period of Earth history. Will we be able to sustain our dominance, continue to grow, or exist as a subdominant species? What factors will determine our fate? The answer definitely involves how we can sustain a clean and productive environment.

References

1. Maton, A., et al. 1993. *Dynamic Earth*, 2nd ed. Englewood Cliffs, NJ: Prentice-Hall.
2. Weinstein, M. P. and R. E. Turner, eds. 2012. *Sustainability Science: The Emerging Paradigm and the Urban Environment.* New York: Springer.
3. Knoll, A. H. 2003. *Life on a Young Planet: The First Three Billion Years of Evolution on Earth*. Princeton, NJ: Princeton University Press.
4. Gordon, M. S. and E. C. Olson. 1995. *Invasions of the Land: The Transitions of Organisms from Aquatic to Terrestrial Life.* New York: Columbia University Press.
5. Wood, B. A. 2011. *Human Evolution: A Brief Insight.* New York: Sterling.

2 Scientific Concepts

Scientific Method

Science is the most democratic discipline on the planet. Scientists have no dogma. They have no need to keep sacred documents or to appoint judges to decide what to believe. Scientists merely have a method, which can be stated—simply: Accept as true only what can be demonstrated.[1] Concepts that cannot be demonstrated fall outside the responsibility of scientists. Concepts that fall outside the realm of science are other disciplines, which look at the world differently. Scientists do not call other disciplines false, unless their practitioners imply that they use the *scientific method* when they do not.

What is meant by *demonstrated*? Another word is *tested*. Scientists start out with some observations in nature and form a hypothesis: a statement that can be tested by experiment, by evaluating a series of observations or measurements, and then having the results peer reviewed by other scientists. An example of a simple hypothesis is: *Water flows downhill.* Another way of stating this hypothesis is: *If water falls on an unobstructed slope on Earth, it will subsequently flow to lower elevations.* Notice, the hypothesis is quite specific. If this hypothesis were tested in a weightless environment, or if the slope had a dam or pond to impede the flow of water, or if the slope was so gradual that the flow would be imperceptible, then the test would not have a clear outcome. A so-called *null hypothesis* would be opposite to what might be expected—that is, water would flow uphill, or would not flow at all. We could set up a testing plan using an inclined plane and a beaker of water.

NOW STOP! What would you expect? If you said this would be a waste of time you might be correct. Intuitively, water will flow downhill and the experiment will show the same result the first time, the second time, and the thousandth time. You can call in other scientists (peer review) and have them confirm your results. They could try this in their own labs over and over again. Okay, the point is made.

Scientists call the *if* part of the hypothesis (water falls on the slope) the *independent variable*, and the *then* part of the hypothesis (flows downhill) the *dependent variable*. Flowing downhill is dependent on the pouring of the water, which is done independently. Note, too, that the cause of the water flowing downhill is not explained fully by the experiment. Gravity is the cause, which pulls objects on Earth closer to each other. Actually, the water is pulling on the Earth as well, but with less influence because of the smaller mass of water with respect to the Earth. It is like the apple falling from the tree.

Although we have a correlation between the slope and the direction the water will flow, there is no *cause and effect* shown—just influence of the slope on the path the water will take. To show cause and effect, we should run the experiment in outer space, where there is weightlessness to see that the slope has no influence as water does not flow down the slope. Without gravity, we would see that globules of water will float or bounce around helter-skelter completely independent of the slope and influence of the gravitational pull of the Earth. But wait! Have you seen how gullible you are, believing our description of the experiments without seeing the results for yourself? To be scientific, *you* must test the hypothesis.

Hypotheses must be able to be tested. An idea that cannot be tested is: *When you die, then you will have life after death*. This is not a scientific hypothesis, because it cannot be tested. Some anecdotal evidence you might hear on late-night radio shows, about dying and coming back to life would not withstand peer review by scientists. The call-in radio guests' stories cannot be tested and peer reviewed. Would you as part of a large group volunteer to be put to death with the hope of being revived and interviewed by a panel of scientists? Just kidding!

When a series of hypotheses that are related to similar natural phenomena are supported by extensive testing and peer reviews, a theory may be developed. The word *theory* has much more certainty in science than in the common vernacular. The dismissive comment "Oh, it's only a theory!" denigrates the validity of many scientific theories. Yet, theories are more certain than hypotheses. A scientific conclusion that has more certainty than a theory is a law. Scientific laws, such as Newton's laws of motion, the laws of conservation of matter and energy, and the ideal gas laws, have not only been tested beyond the theory stage, but have mathematical predictability.

Nevertheless, we must keep in mind that even scientific laws have their weaknesses. Hypotheses, theories, and laws can be changed, discredited, and reversed when new evidence comes along to discredit old ideas. The ancient Greeks believed that the Earth was round and was not the center of the universe. The Dark Ages dogmatically countermanded those beliefs. Galileo had to disclaim his own scientific conclusions to save his life. He would have been put to death if he espoused his conclusions that the Earth revolved around

FIGURE 2-1 Big Bang theory.

the Sun (which it does, of course). Even Newton's laws of motion and the conservation laws have been modified by Albert Einstein, with respect to, among other things, matter and energy transformations.

The Big Bang

Scientists have observed astronomically that the universe is expanding. They theorize from current measurements of rates of expansion that this process started about 13.7 billion years ago, originating from a small, dense cloud of energy and simple atoms (e.g., hydrogen, helium). The start of this expansion was cataclysmic, and is called the *Big Bang* **(Figure 2-1)**.[2]

This cloud began to expand and form galaxies of stars undergoing energy and matter transformations. Subatomic particles of simple atoms were combined by nuclear fusion in the stars to form an array of more complex atoms, the so-called elements of the *Periodic Table.* Although the total quantity of matter and energy in the universe is constant, transformations from one form of energy to another render the available energy (ability to do work) less and less useable (*entropy*) over time. Eventually (in billions of years into the future), energy will be so dissipated that it would take another Big Bang to rejuvenate useful energy sources. In the meantime, we can use various forms of nonrenewable and renewable energy sources while they are sustainable.

Evolution

The theory of *evolution*, now an accepted concept in the scientific world, is an important topic in environmental science.[3] Although much misunderstood by nonscientists, evolution accounts for the myriad of life forms on the planet and explains their diversity, interactions, and history.

Organisms are influenced by *abiotic factors*. These include temperature, moisture, salinity (for marine environments), inorganic nutrients, space to live, and Earth processes. These influences determine where and how an organism survives. These factors act not only on the individual organism, but on all individuals of the organism's species, its *population*. Some of the influences are critical to its survival, namely limiting factors that the species cannot do without. Scarcity or absence of water, nutrients, and certain foods may limit the ability of the species to thrive or survive. Populations thrive when these factors are optimum, or lying within a range of tolerance (narrow or broad) that are beneficial to the needs of the majority of the individuals. If the environment is outside the range of tolerance, the population dwindles, could die out, or is forced to migrate to areas having their optimum range of tolerance.

Organisms are also influenced by biotic factors, namely organism interactions. These include relationships among different organisms, such as predator–prey relationships, *symbiosis* (living together with or without mutual benefit), disease, and competition for food or mates.

Evolution acts at the population level. The assumption that evolution operates at the individual organism level is the source of much misunderstanding espoused by doubters of the concept of evolution. *Populations* are composed of organisms of the same species, but members have slightly different ranges of size **(Figure 2-2)**, appearance, and tolerances. The range of tolerance of the members of the population could be graphed as a *bell-shaped distribution curve.* **(Figure 2-3)**.

The vertical axis represents the number of individuals of the species and the horizontal axis represents the *range of tolerance* of a specific factor. Let us consider an example of how the factor of temperature affects a certain species of moth.[4] The highest temperatures fall in the tail (farthest right); the lowest in the other tail (farthest left). The greatest number of this species (highest in the middle) flourishes at a temperature of 75° F (24° C), so at this temperature we have the largest population. The population falls off in higher and lower temperatures. At the "tails" of the curve, there are very few moths of this species at temperatures around 40° F (4.5° C) and around 110° F (43° C) and no moths lower or higher. Those in the tails are rather rare and strange to the species. You might call them mutants (or runts of the litter). They can exist at temperatures where most others of their population cannot.

FIGURE 2-2 Same species, different size and appearance

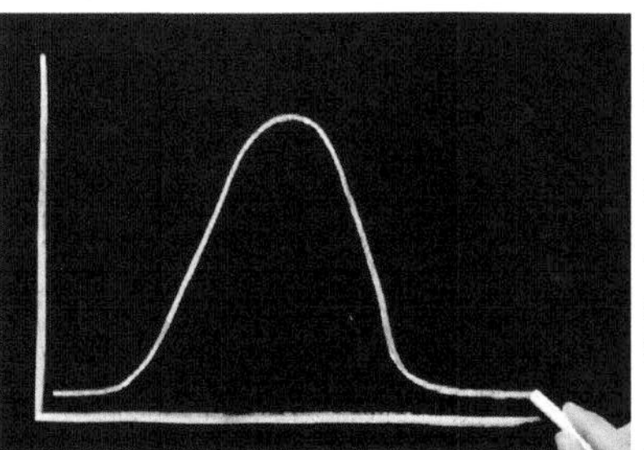

FIGURE 2-3 Bell-shaped distribution curve

Now, let us say the climate changes. It gets much warmer. Later generations of these moths at the upper end of the temperature range are favored for survival. Later generations at the lower end of the temperature range are in disfavor for survival. Thereby, the bell-shaped curve maximum shifts towards the warmer temperatures, with maximum populations at those temperatures that once were in the upper end tail of the curve. Moths now in the majority produce more offspring that have this warmer temperature tolerance.

What causes this temperature tolerance? Genetic makeup causes this trait. Consequently, the gene pool becomes enriched in moths that favor the warmer temperatures.

Another trait that might become more or less prevalent (for other reasons) is color. Let us say a population of moths of light and dark colors lives in a forest of white-bark birch trees. Those individuals having dark colors that land on white bark are more likely to be spotted by insect-eating birds and be eliminated. The light colored moths are in protective camouflage and survive. The bell-shaped curve would then shift so that the majority has light colors. But if industrial pollution darkens all the tree barks, then the change would shift in favor of dark-colored moths.

These examples of traits that produce dynamic changes in populations demonstrate the process called *natural selection*. This process was proposed by Charles Darwin in his book *Origin of Species* published in 1859.[5]

The dynamic changes described in our examples could happen in a few years. More significant changes in population traits would require more time and more cycles of change. Changes in populations that require millions of years to develop and occur over thousands of generations could affect more than color or temperature tolerance. In each generation, end members in the tails of the distributions could be quite different from the original populations. If they are so different and isolated (geographically or genetically) from the main populations repeatedly, then they may lose the ability to interbreed with those of their species. At that point, we say that speciation has occurred.

© Alan Jacobs

FIGURE 2-4 Ape and human skeletons, American Museum of Natural History

Isolation could occur by the formation of a geographic barrier (e.g., lava flow, major canyon, continental separation, land bridge flooded) and separated habitats of this species experiencing different environments (e.g., warmer, colder, different predators, different food availability). Genetic barriers could also form, preventing mates from producing fertile offspring that would carry on different traits. The result would be two different species (**Figure 2-4**).[6]

Evolution is an environmental service. With natural selection, populations of organisms adjust to changing environmental conditions by readjusting species population environmental tolerances to adapt to these changes. Organisms within a population (of each specific species) will be most abundant at optimum environmental conditions and least abundant at the limits of the ranges of these conditions. Yet, when environments change, the elements of the populations at the range limits (mutants) may be more successful in future generations of the species and will pass on their genetic traits to the exclusion of the rest of the population. Thus the mutants, so to speak, become the majority.

Individuals that may have been in the majority are now in the minority. Under new environmental conditions, the species thrives thanks to the environmental tolerances of the new majority. Thus, the species is sustained and life goes on.

Some species cannot keep pace with the changing environment. Mutations may not be fast enough to produce offspring that can prosper in the new environments. Consequently, a species can become extinct. Numerous species have become extinct during Earth history. It is Nature's way of providing niches for newly speciated populations that may be better able to survive in the future.[7]

SUCCESSION

The soil, the land surface itself, the climate, and the available moisture can change on a dynamic Earth. When a *catastrophic event* occurs, the ecosystem adjusts to the changes in a series of "repairs" **(Figure 2-5)**. The repairs can create intermediate environments, themselves unstable, which can be replaced successively until a dynamic equilibrium is reached.

Image © Roberto Caucino, 2014. Used under license from Shutterstock, Inc.

FIGURE 2-5 Ravaged landscape in need of repair.

For example, a volcano erupts and covers all productive soils with hard lava rock. The soil and vegetation below is burned and buried. Wildlife dies or they emigrate to areas not affected by the volcanic eruption. Slowly, the lava rock cools and becomes weathered, primitive soils form, and pioneer vegetation establishes itself and forms local microclimates. These microclimates welcome other plants in a series of stages. Animals find temporary habitats here. Both plants and animals are replaced as the biome recovers.

If recovery leads from a completely barren environment,[8] this *primary succession* can be slow. If the disturbance is only partial, then the ensuing *secondary succession* can proceed more rapidly. What is the end point to the series of changes in this succession? Theoretically, the land will return to the environmental setting (*biome*) from which it had been in equilibrium prior to the disturbance. The community to be reestablished is called a *climax community*. The Earth is dynamic, so there may be differences between the original community and the climax community.

References

1. Gimbel, S., ed. 2011. *Exploring the Scientific Method: Cases and Questions.* Chicago: University of Chicago Press.
2. J. R. O'Connell and A. L. Hale, eds. 2012. *The Big Bang: Theory, Assumptions, and Problems.* Hauppauge, N.Y.: Nova Science Publishers.
3. Hull, D. L. 1983. *Darwin and His Critics: The Reception of Darwin's Theory of Evolution by the Scientific Community.* Chicago: University of Chicago Press.
4. Sagan, C. 1977. *The Dragons of Eden: Speculations on the Evolution of Human Intelligence.* New York: Random House.
5. De Beer, G. 1964. *Charles Darwin: Evolution by Natural Selection.* Garden City: Doubleday.
6. Boggs, C. L. and J. Roughgarden. 2000. "Biodiversity: Result to Speciation and Extinction." Chap. 17. *In Earth Systems: Processes and Issues edited by* Ernst, W. G. New York: Cambridge University Press.
7. Landweb, L. F. 1999. *Genetics and the Extinction of Species: DNA and the Conservation of Biodiversity.* Princeton, N.J.: Princeton University Press.
8. Enger, E. D. and B. F. Smith. 2008. *Environmental Science: A Study of Interrelationships,* 12th ed. Boston: McGraw-Hill.

3 Earth Spheres

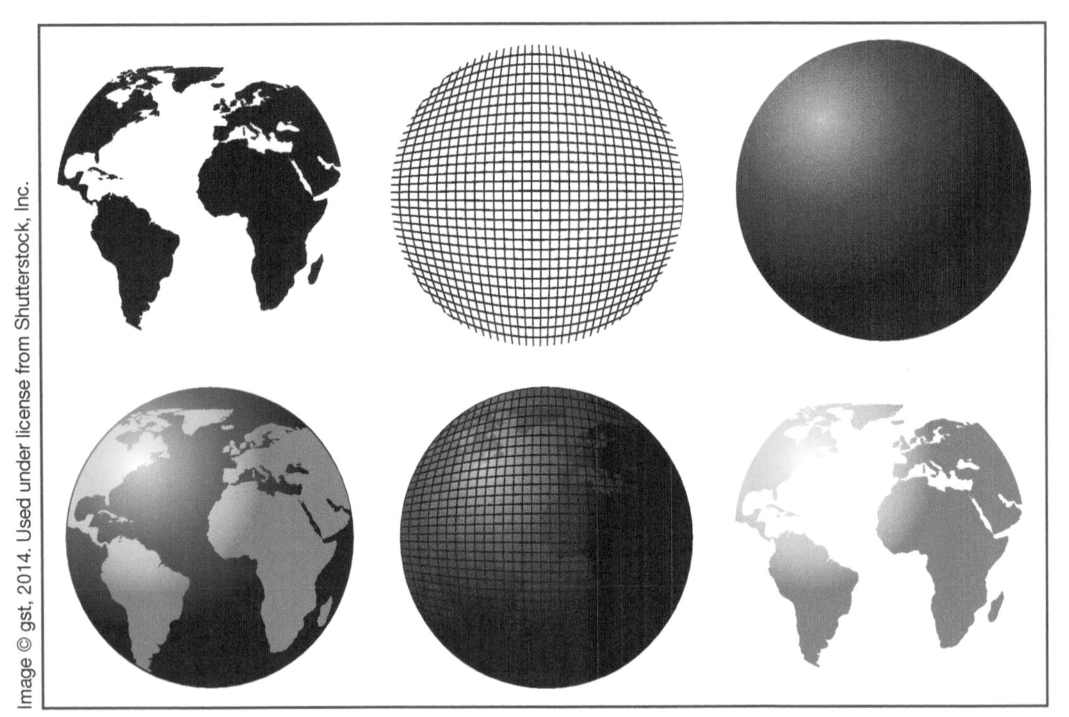

The Earth is studied by scientists of various disciplines. We can divide the Earth into *spheres* based on which scientists study different parts of the Earth.[1] These spheres are not geometric shapes, but rather spheres of study. Geologists study nonliving spheres, including solid and molten rocks (*geosphere*), water and glaciers (*hydrosphere*), and air and wind currents (*atmosphere*), collectively called the *geosphere*. Biologists study life forms of the *biosphere*.

The Earth, a planet in our solar system, is also studied by astronomers. Our *solar system* is situated among star clusters, called *galaxies*. Our solar system is in the *Milky Way Galaxy*. Finally, galaxies constitute the matter of the universe. Astronomers study the *exosphere*.

Environmental science incorporates most of these spheres of scientific study, but concentrates on those which affect Earth's environments. Therefore, we will forgo the details of subatomic particles and distant galaxies to the quantum physicists and the astrophysicists, respectively. Consequently, our environmental sphere of study extends from atoms to bodies in space in our corner of the solar system.

Geosphere

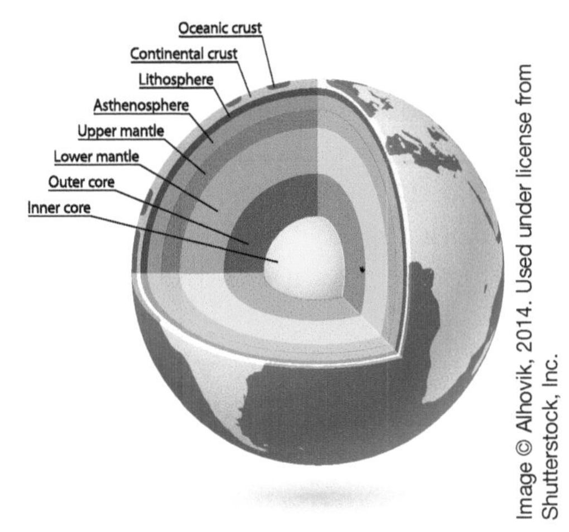

FIGURE 3-1 Geosphere: layers of the Earth

The *geosphere* **(Figure 3-1)** comprises all solid rock, molten rock, sediment, and soil on Earth.[1] This sphere is stratified into concentric layers. The least dense, outer layer is called the *crust*. It can be divided into low-density rock (granitic) that underlies the continents and higher-density rock (basaltic) that underlies the oceans. Because the *basaltic crust* is denser than the *granitic crust*, it lies lower in elevation and is inundated by most of the oceanic waters. The granitic crust "floats" higher in elevation to raise the crust above sea level and afford a so-called *terrestrial environment* for us land creatures. The remainder of the oceanic waters (of the *hydrosphere*) extends onto the continental margins.

The crust is underlain by rocks having a greater density than the densest crustal material. Originally the outer layers of the geosphere were divided into three: crust, *mantle*, and core. As a result of *tectonic plate theory* (see below), the crust and the upper, rigid part of the mantle are combined and called the *lithosphere*.

The rocks below the crust constitute the *mantle*. The upper part of the mantle and the crust (lithosphere) move in unison as rigid plates over an underlying plastic part of the mantle called the *asthenosphere*. The mantle contains hot spots that cause crustal melting, new crust formation, volcanic eruptions, rigid plate movements (by heat convection), and crustal destruction.

The mantle is underlain by a two-layer *core* of extremely dense material. The outer core is liquid and the inner core is solid. We know this by the way different earthquake waves bend and are absorbed as they pass through the core. The outer core is thought to be responsible for the Earth's magnetic field.

At the upper part of the crust, we find *sediment* and *soil*, which is derived from crustal rock. These terms have different meanings in different disciplines. For environmental science studies, both sediment and soil are unconsolidated particles of rock and other matter. In this context, sediment is the weathered products of rocks that are being transported for eventual deposition and reformation into solid rock. Soil is sediment that is not necessarily destined to become rock.

Hydrosphere

The *hydrosphere* comprises all water on Earth[2]—in oceans, lakes, streams, aquifers, precipitation, glaciers, clouds, soil, wetlands, living creatures, and waters of hydration in minerals **(Figure 3-2)**. This is a closed system; i.e., no water enters or leaves the planet except for a few additions of ice from comets that fall to Earth. Water is recycled, and in the process is naturally cleansed by evaporation and filtration, and used to transport sediment, weather and erode rock, and transfer energy (heat and mechanical energy).

FIGURE 3-2 Hydrosphere

Atmosphere

The *atmosphere* is composed of gases, dust, and water vapor. Its concentration is greatest near the surface and becomes thinner upwards until a vacuum is reached beyond about 250 miles in altitude.[3,4] Layers of the atmosphere are distinguished based on changes in temperature fluctuations with altitude **(Figure 3-3)**.

- *Troposphere* : 0–18 km (~0–11 miles)—Breathing zone, weather, smog—Temperature decreases with height
- *Stratosphere*: 18–50 km (~11–31 miles)—Ozone "layer"—filters UV radiation—Temperature increases with height
- *Mesosphere*: 50–80 km (~31–50 miles)—Meteorites disintegrate (aircraft fly below this layer)—Temperature decreases with height
- *Thermosphere*: 80–100+ km (~50 to >60 miles)—Ionizing air particles—Temperature increases with height; heat content falls.
- *Exosphere*: >100 km (60 miles)

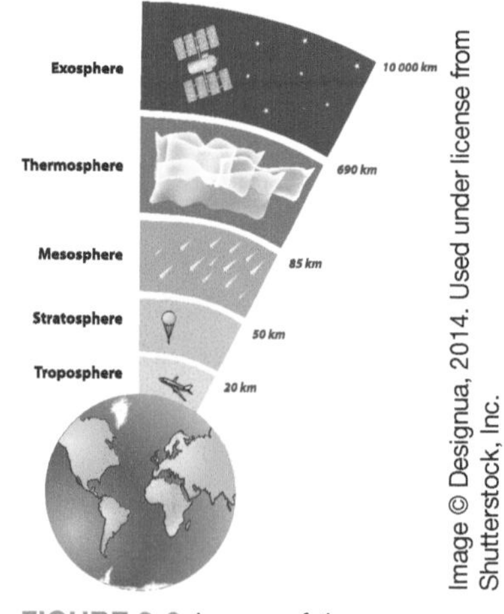

FIGURE 3-3 Layers of the atmosphere

Biosphere

The *biosphere* comprises all living creatures **(Figure 3-4)**.[5] Life in this sense is defined as the ability of a carbon-based unit to replicate and sustain itself. Life forms consist of plants, animals, and other organisms. Environmental scientists like to define life forms according to their role or *niche*. Therefore, they may be *producers* (plants), *consumers* (animals), or *decomposers* (e.g., bacteria, fungi).

The *dynamism of* the Earth is manifested in the lithosphere (earthquakes, plate tectonics, volcanism), the hydrosphere (floods, glaciers, tsunami), the atmosphere (weather phenomena, climate change), and the biosphere (extinction, migrations, succession, evolution).

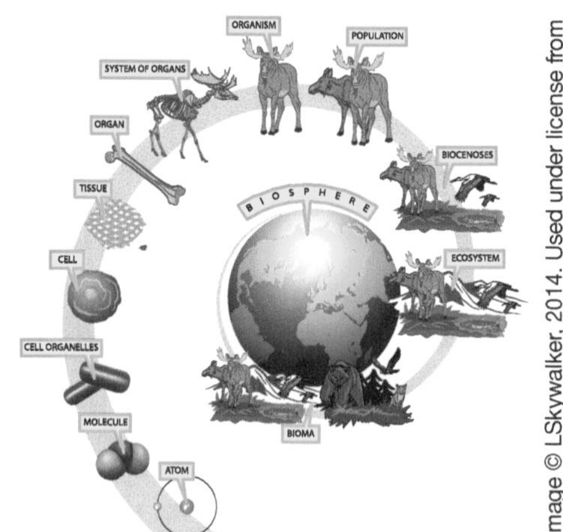

FIGURE 3-4 Life forms in the biosphere

References

1. Jacobs, A. 2013. *Need to Know in Environmental Science*. San Diego, CA: Cognella.
2. Plummer, C. C. et al. 2008. *Physical Geology*, 13th ed. New York: McGraw-Hill.
3. Linde, B. M. 2005. *Water on Earth*. Pelham: Benchmark Educ. Co.
4. Wells, N. 2012. *The Atmosphere and Ocean: A Physical Introduction*. Hoboken, NJ: Wiley.
5. Enger, E. D. and B. F. Smith. 2008. *Environmental Sciences: A Study of Interrelationships, 12th ed*. New York: McGraw Hill.

UNIT 2 Science and the Environment

All the sciences provide a foundation for environmental science, so it is instructive to review the basic information of each science and relate it to the environment. Physics and astronomy deal with matter and energy. Chemistry provides details of different forms of matter and the energy required for matter to interact. Geology provides details of more complex systems of matter on Earth and natural processes that use available energy. Biology deals with living forms of matter and the flow of energy through living beings. This is probably an oversimplification, as natural complications make it difficult to compartmentalize the various sciences. Overlapping relationships among the sciences, therefore, result in subdisciplines of each of these four basic sciences called, for example: geo-, astro-, and biophysics; geo- and biochemistry; physical chemistry; and paleontology. Chapter 3 on Earth spheres emphasized this interdisciplinary aspect of science, in general, and environmental science, in particular.

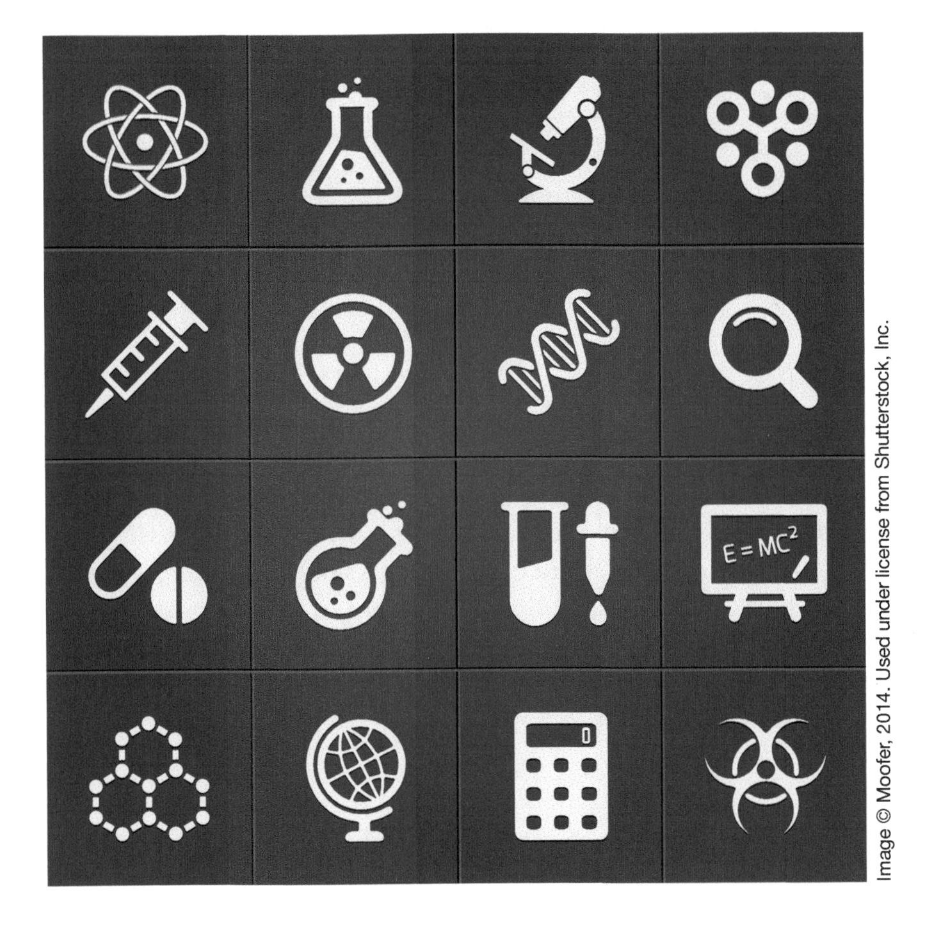

4 Physics–Astronomy and the Environment

"We cannot solve our problems with the same thinking we used when we created them.

Albert Einstein

Matter

Matter can exist in three physical states: solids, liquids, and gases. Matter has physical characteristics, which are important in discussions of our planet's environment, e.g. *mass* (amount of matter), *weight* (force from gravity), *volume* (space occupied), *density* (mass per unit volume), *dimensions* (size measurements), *strengths—tensile* (resistance to being pulled apart), *compressive* (resistance to being crushed), *sheer* (resistance to being twisted), *color—hue* (part of the visible light spectrum, e.g., blue, red, green), *value* (light or dark), *and saturation* (intensity of the color), and abilities to store and transmit energy. Environmental science does not need to consider antimatter, which seems not to exist on or affect our environment on Earth.

Atomic Structure

Atoms and their subatomic particles are the smallest building blocks of the universe.[1] The number and kind of subatomic particles in each atom (protons, neutrons, and electrons) determine their identity and properties. For example, helium has two *protons*, two *neutrons*, and two *electrons* (**Figure 4-1**).

There are ninety different naturally-occurring elements, ranging from hydrogen to uranium, which are shown on a periodic table (**Figure 4-2**) of elements,[2] from atomic numbers 1 through 92. The two extra elements in this range (technetium and promethium) have no stable isotopes, and therefore are not considered naturally occurring. Examples of other elements are iron, nitrogen, lead, sulfur, oxygen, and carbon. The periodic table also contains more than a dozen elements that have been artificially produced.

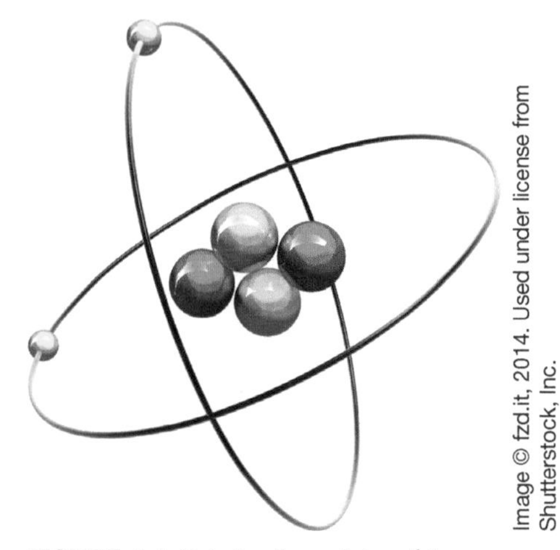

FIGURE 4-1 Subatomic particles of the helium atom

Each element has a defined number of subatomic particles in its atoms called *protons*. The proton has a positive charge and a unit mass, and resides in the nucleus of the atom. All atoms of each element have a fixed number of protons—for example: iron 26, hydrogen 1, nitrogen 7, lead 82, sulfur 16, oxygen 8, and carbon 6—in each of their nuclei. The nucleus of the atom of any of the elements can have one or more additional particles, called *neutrons*. The neutron has a neutral charge (no charge at all) and a unit mass, and also resides in the nucleus of the atom. The number of neutrons in the nucleus does not change the atom into another element but adds weight or mass to the atom. *Naturally-occurring hydrogen* is the only element that can have no neutrons. When an element can exist in nature with a varying number of neutrons it forms an isotope of that element. Even hydrogen can have one or two neutrons in addition to its one proton, which forms isotopes of hydrogen called *deuterium* and *tritium*, respectively. Carbon has six protons and six, seven, or eight neutrons.

Surrounding the nucleus of protons and neutrons are subatomic particles called *electrons*. They are found in groups in so-called *electron shells* of successively greater energy states and number at greater distances from the nucleus. Electrons have a negative charge and hardly any weight at all. In an electrically neutral atom, the number of

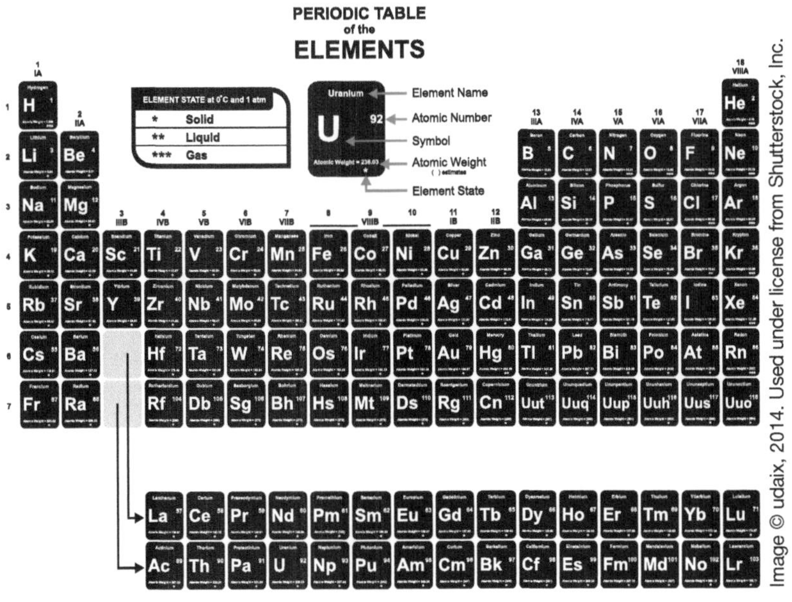

FIGURE 4-2 Periodic table of the elements

electrons equals the number of protons. When the atom has a net charge—positive when it loses electrons and negative when it gains electrons, it is called an *ion*. The electrons can be shared or lost to other ions causing the resulting electrical charge to bind the atoms together forming compounds. Each unit of the compound is a *molecule*.

Energy

Energy is the ability to do work **(Figure 4-3)**.[3] Most of our energy comes from the thermonuclear reactions in the Sun, the star of our solar system. Some *energy on* Earth is also provided by mineral radioactivity within the interior of the Earth (fuel for volcanoes and crustal plate movements). The Sun has been sending energy to the Earth starting more than 4 billion years ago and will continue to do this for another 5 billion years, at least. We owe our existence to the flow of this energy through the biosphere, geosphere, hydrosphere, and atmosphere.

Energy at rest, like that big boulder on the mountain ready to roll down by the force of gravity, is *potential energy*. It has the potential to crash and erode the mountainside and break up into smaller pieces until it comes to rest at the base of the slope. When that boulder is in motion, the potential energy is transformed into *kinetic energy* (energy of motion). The work in moving the boulder down the mountain (by the force of gravity) also—through friction between the boulder, the mountainside and the surrounding air—produces heat that is dissipated during its descent.

Was the energy used up? No, energy can neither be created nor destroyed. Unfortunately, the universe has a finite amount of energy. With each transformation, energy becomes less able to be reused. This property of energy is called *entropy*. The frictional forces produce heat that is dissipated. If you use fuel to accelerate your automobile and then apply the brakes, the stopping of the car produces heat in the brake linings of the wheels that dissipates into the atmosphere. If you put a vessel of water adjacent to your brake linings, you could boil water for tea—but that is probably not in your plans.

What was the original source of all this energy? Astrophysicists theorize that energy was created during the Big Bang, at the formation of the universe more than 13 billion years ago. Since then, energy has been used and heat has been dissipated. There will be enough energy to last many lifetimes. What we are working on is efficient, economical, sustainable, and environmentally friendly ways of tapping the energy provided to us by the Sun, the Earth's interior, and gravitational forces between the Earth and other bodies in our solar system.

Solar energy through direct solar radiation has been stored for us, thanks to plants and microscopic animals. On a daily basis, through the process of *photosynthesis*,[4,5] plants take water and carbon dioxide and convert them chemically into carbohydrates and oxygen with the help of solar radiation and a green catalyst called *chlorophyll*. The energy of this transformation is stored in the chemical bonds of the carbohydrate, glucose. This is another example of potential energy. When we eat carbohydrates or tissues from animals that have eaten the

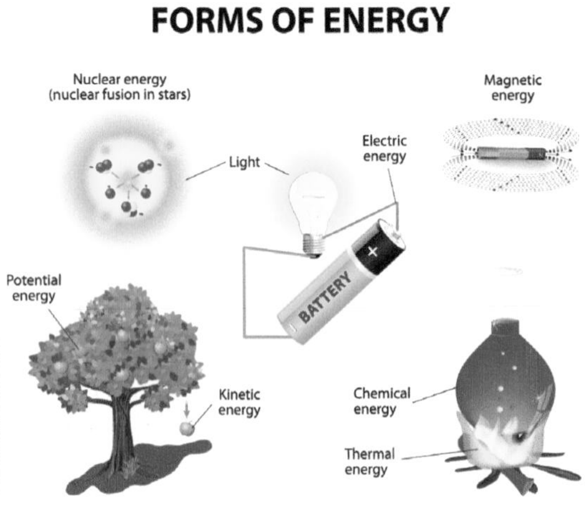

FORMS OF ENERGY

Nuclear energy (nuclear fusion in stars)
Magnetic energy
Electric energy
Light
Potential energy
BATTERY
Kinetic energy
Chemical energy
Thermal energy

FIGURE 4-3 Forms of energy

carbohydrates together with breathing the oxygen, we break down the carbohydrates and change them back into carbon dioxide and water and release the energy from the chemical bonds—a process called *respiration.*[6] We use this energy for growth, internal bodily functions, and activity. As we use this energy, we transform the potential energy we store in our bodies to kinetic energy to perform life's activities.

On a long-term basis, the Earth has transformed and stored energy containing products derived from excess plant and animal tissues into fossil fuel deposits (wood, plant wastes, peat, coal, oil, and natural gas). This is another example of potential energy. We have extracted these fuels and continue to deplete the supply of those fuels that are nonsustainable. Coal, oil, and natural gas are nonsustainable because it would take millions of years for Nature to form them again.

We have learned to exchange different kinds of energy for our needs. *Mechanical energy*, which directly moves objects, can be obtained or transformed into heat energy, phase-change energy, electromagnetic energy, chemical energy, electrical energy, and nuclear energy. For example, to generate electrical energy to store in batteries, we might burn fossil fuel (releasing chemical energy), to heat water (heat energy), to create steam (phase-change energy), to turn turbines (mechanical energy), to get electrons moving in the turbine coils (electrical energy), and to charge batteries (chemical energy). That electrical energy generated could also run furnaces (heat energy), power electrical tools (mechanical energy), and light up homes or run computers (electromagnetic energy). Entropy is present at every energy transformation, so it is wise to capture and use the dissipated heat before it is rendered unusable.

Aside from plants directly capturing the radiation from the *Sun* and converting it into energy-storing plant products, the unequal heating of the Earth's surface, indirectly transfers some of the Sun's radiant energy to the atmosphere, where warm, rising air currents and cool, falling air currents form circulation cells. These circulation cells power the weather systems, wind currents, and most of the ocean currents and waves, and effect evaporation and global warming and cooling.[7]

Other bodies in our solar system also add energy to the planet by gravitational attraction, producing Earth waves called *tides*.

With all these sources of energy, you might ask, why is there an energy crisis? We hope you can answer this question by reading further.

The Sun

The Sun **(Figure 4-4)**, our nearest star, is on average 93 million miles (about 150 million km) away. Its energy, traveling as electromagnetic waves at a rate of 186,000 miles per second (300,000 km/second), reaches the Earth in about 8 minutes.[8,9] The waves can pass through a vacuum for most of the way and many frequencies of these waves can also penetrate our atmosphere.

FIGURE 4-4 The Sun

The Sun is our powerhouse, which provides and has provided energy (stored in *fossil fuels*) that we use every day. *Thermonuclear fusion reactions* on the Sun combine hydrogen to form helium from huge gravitational compression forces, which releases electromagnetic energy in all directions. In the path of this energy is Earth.

An immediate benefit from this input of energy is a one-percent capture of incoming solar radiation by plants, which, through photosynthesis, changes carbon dioxide and water into oxygen and glucose sugar. The energy from this process is stored in the chemical bonds of the sugar, holding atoms of oxygen, carbon, and hydrogen. Plants use a catalyst, a substance that helps in this conversion at lower energy levels, called *chlorophyll.* This catalyst is green, not because plants prefer green light energy, but conversely because they "prefer" the blue and red parts of the light-energy spectrum to perform photosynthesis. Their green color reflects the green light away, so, in effect, the plant removes less efficient energy from the photosynthetic process.

The plant material is used as food for animals and decomposers. Dead plant material and microscopic animals that feed on plant material have been preserved in the crust over millions, tens of millions, and a couple of hundred million years, with much of their stored energy intact. After burial of plant material, heat, and chemical reactions over tens to a couple of hundred million years, coal was formed, a common fossil fuel. The preserved microscopic animals and plants through similar processes over tens of millions of years have formed oil and natural gas, other common fossil fuels. All this energy was derived from the Sun, through photosynthesis.

The Sun is also the energizer for additional sources of useful energy. Unequal warming of the atmosphere causes warm air to rise and cool air to fall. Together with the rotation of the Earth on its axis, these air masses move and transmit mechanical energy which can turn wind turbines to generate electrical energy.

The Sun, also aided by the gravitational pull of the Earth, energizes the hydrologic cycle, evaporating water that rises into clouds, falls as precipitation, purifying water through evaporation and filtration, and results in storing water at high elevations to be used to turn hydroelectric turbines, which generate electricity.

The Sun also exerts its pull on the Earth and moves ocean water in the form of tides. Our Moon is much closer, though smaller, and has a greater influence on tides. Both the Sun and Moon, however, move large quantities of water onto the continental shelves and into embayments. In bays, the movement of the water (periodically daily or twice a day) can be harnessed by tidal turbines placed to capture the rush of the tides (in or out) and convert this tidal energy into electricity.

Direct rays of the Sun can be captured by *solar cell panels,* which convert solar radiation directly into electricity. Or the Sun can heat up fluids that vaporize and the vapors can turn turbines to generate electricity. Furthermore, the Sun can heat up ceramic tiles during the day and give off heat energy at night. This solar heat energy can be used directly without the need to convert the heat to electrical energy, and then use electric-powered heaters to turn the electricity back into heat.

Some *forms of* solar energy are more sustainable that others. Because we have a finite supply of fossil fuels in the crust, our growing population, technology, and thirst for energy will eventually exhaust our supply of coal, oil, and natural gas. It is true that we keep extending this supply with new finds and, to a minor degree, with reuse of some cooking oils to power vehicles, but the demand for fossil fuels will eventually surpass its supply. Conservative estimates range from one hundred to a few hundred years. Even optimistic estimates of fossil-fuel reserves might stretch this run-out time to a thousand years.

Direct use of photosynthetic production can last as long as the Sun provides us with energy. The good news is that solar radiation delivered to Earth (with free shipping) should last another five billion years. Through the harvest of annual crops of corn, sugar cane, and other plants (even algae) biofuels are more sustainable than fossil fuels. One limitation of this supply would be depletion of soil components that add to the nutrients that fuel plants need. Also, the plants used for biofuels compete with crops used for our food supply. Do we use corn for cattle or for ethanol in our automobile tanks? Can we recycle soil nutrients fast enough to keep up crop production for both food and fuel?

Direct use of *nonphotosynthetic sources of* solar energy (e.g., solar panels, wind turbines, hydroelectric power, ceramic tiles, tidal power plants) is also more sustainable than fossil fuels. Unfortunately, the technology, economics, and politics (with lobbying run rampant) needed to make these forms of energy replace fossil fuels is a difficult societal challenge. We are making inroads, but have a long way to go—e.g., hybrid and total electric cars, wind turbine and solar panel installations, eco-construction of new homes, tidal power plants. Aside from geothermal energy, all other sources of sustainable energy are directly or indirectly tied to energy from the Sun.

The Sun also emits dangerous radiation in the form of *ultraviolet* (UV) and *infrared rays*. We must regulate our exposure to these rays to reduce our absorption of these rays. Consequently, we must avoid sunbathing without UV-ray protection, avoid tanning salons, and discourage the use of chemicals that reduce ozone levels in the stratosphere (e.g., *chlorofluorocarbons*). Other effects of the Sun on our environment include evaporation of water supply reservoirs, unequal heating of the Earth, resulting in severe weather conditions, climate change from fluctuating emission of solar radiation and greenhouse effects, and droughts producing forest and grassland fires.

We are indeed fortunate that the Sun is at a safe distance from the standpoint of gravitation pull and also for lower radiation and protection from thermonuclear explosions. We are also fortunate that the Sun is near enough to provide us with light and heat to sustain life on this planet.

The Moon

We have one moon (the Moon) that orbits the Earth every $27\frac{1}{3}$ days and is approximately 225,000 miles (363,000 km) from the Earth **(Figure 4-5).**[10] Some aspects of life on Earth, like biorhythms, may be affected by our Moon. Although folklore adds many other lunar influences, one sure thing is the effect of the gravitational attraction on near-shore waters and its resulting tidal effects. High tide and low tide are documented occurrences that influence shipping, the life of marine organisms in the tidal zone, salinity of the water at the mouths of streams emptying into the ocean, erosion of coastal cliffs and beaches, and as mentioned under the joint influence of the Moon and Sun, electrical power generation using tidal power plants.

FIGURE 4-5 The Moon

Why does the Moon influence the tides more than the Sun does? The answer is, in part, the *inverse-square law,* acting on the gravitational pull of bodies in space. Simply stated, for every unit of distance closer to an object, its pull is equivalent to the square of that distance. Consequently, if we are one-half the distance away, the gravitational pull is four times as great. If we are one-third the distance away, the gravitational pull is nine times as great. So, it would seem, that the inverse-square law would make the Sun, from a distance of 93 million miles, have a much lower pull than if it were as close as the Moon. But even at 400 times that distance that solar tidal force would still be about 175 times greater than the lunar tidal force. Another factor, however, is also involved. Tides are caused by the difference in gravitational pull across the Earth. The Sun is so far away compared with the diameter of the Earth, that the difference in tidal pull across the Earth is lower than expected. So it turns out that the tidal pull of the Sun is only 44 percent as strong as the tidal pull of the Moon.

Because the Moon is orbiting the Earth and the Earth is rotating on its axis, the *tidal effects of* the Sun's and Moon's gravitational forces can change with respect to their positions. Forces can be additive and affect different sides of the Earth at the same time. Consequently, we have strong tides (so-called *spring tides*) when the Sun and Moon are pulling in the same direction, or weaker tides (so-called *neap tides*) when they are pulling at right angles.

Tidal effects are also local, as different locations on Earth are facing the Sun and/or Moon at different times of day. The shapes of coastal areas also affect how the tidal pulls affect the flow of water in embayments and up the mouths of rivers that empty into the seas.

Other Bodies in Space

Other than the Sun and Moon, our environmental conditions can be affected by objects in space that fall to the Earth and by nonsolar cosmic radiation. Meteors that fall to the Earth are called *meteorites* **(Figure 4-6)**. They are quite dense and travel at high speeds, and so they can be dangerous. Fortunately, most meteorites do not hit the Earth's surface, because they disintegrate from friction with air molecules after entering the Earth's atmosphere. This disintegration occurs in the mesosphere (31 to 50 miles (50 to 80 km) in altitude). Most of these meteorites are observed by their emission of light during entry as so-called *shooting stars*. There have been infrequent incidences of small meteorites hitting people or larger ones exploding in populated areas—most recently in Russia in February 2013.

FIGURE 4-6 Flare of a falling meteorite

Comets, large chunks of mostly ice, can also be observed in their path around the Sun when coming close to the Earth. At most, if they enter the atmosphere, they may add some water to the hydrosphere.

References

1. Aloian, M. 2008. *Atoms and Molecules: Why Chemistry Matters.* New York: Crabtree.
2. Scerri, E. R. 2011. *The Periodic Table: A Very Short Introduction.* Oxford: Oxford University Press.
3. Raven, P. H., L. R. Berg, and D. M. Hassenzahl. 2010. *Environment.* Hoboken, NJ: Wiley.
4. Ghosh, T. K. 2009. *Energy Resources and Systems.* Dordrecht: Springer.
5. Renger, G., ed. 2008. *Primary Processes of Photosynthesis: Principles and Apparatus.* Cambridge, UK: RSC Publishing.
6. Anderson, D. K. "Mechanics of Respiration." Chap. 31. *In* Banks edited by R. O. *Essentials of Physiology.* Boston: Little, Brown.
7. Barry, R. G. and R. J. Chorley. 2010. *Atmosphere, Weather, and Climate.* New York: Routledge.
8. Dwivedi, B. N., ed. 2003. *Dynamic Sun.* Cambridge: Cambridge University Press.
9. Mullan, D. J. 2010. *Physics of the Sun: A First Course.* Boca Raton: CRC Press.
10. Hicks, T. A. 2010. *Earth and the Moon.* New York: Marshall Cavendish Benchmark.

5 Chemistry and the Environment

With some overlap with physics, chemistry also deals with matter and energy. Its studies differ from physics, because it focuses on the differences in types of matter and the kinds and amounts of energy required separating and combining various types of matter.

The relationship between chemistry and *environmental* science concerns the impact of various chemicals, natural and manufactured, on our health and the environment. Ironically, we are all chemicals. Our environment is composed of chemicals. Some of them are vital to survival, others harmless, and yet others harmful and sometimes fatal.

In environmental science an important subdiscipline of chemistry is called *toxicology (the study of poisons)*. From toxicology we learn that some chemicals are more poisonous than others. We also learn that all chemicals can be poisonous at certain doses. The negative impact of chemicals, therefore, depends on the dose and the strength of those chemicals. Another important consideration is the kinds of exposure to these chemicals. The resulting environmental impact to our health and the environment can be evaluated by comparing risks to these chemical exposures.

A unifying theme and starting point among chemical studies is its use of the *periodic table of the elements.*

The Periodic Table of the Elements

The periodic table **(Figure 5-1)** lists the elements using character symbols derived from their names in various languages.[1] The name for lead in Greek is *plumbum*, represented by the symbol Pb. Therefore, some symbols must be learned for proper identification, such as Sn for tin, K for potassium, and Au for gold. Some of the easier ones for English speakers are C for carbon, H for hydrogen, Ca for calcium, and U for uranium. The element box lists the element symbol, the number of protons called the *atomic number*, and the average atomic weight/mass of the isotopes of each element.

The weight of each atom is the total of the number of protons and neutrons. However, the table lists the average weight of all the isotopes of each element. As some isotopes are more abundant than others, the average weight listed may not be a whole number. For example the atomic weight of hydrogen is listed as 1.008, as there are a small number of deuterium and tritium isotopes in nature that increases the atomic weight above 1.000. The weight of a hydrogen atom without any neutrons is only 1, derived from the weight of the proton.

A *molecule* is a unit of a compound made up of atoms of one or more elements. Molecular compounds of atoms can be formed when metallic and nonmetallic ions combine. The bonds between them are called *ionic*. When atoms share electrons (as among

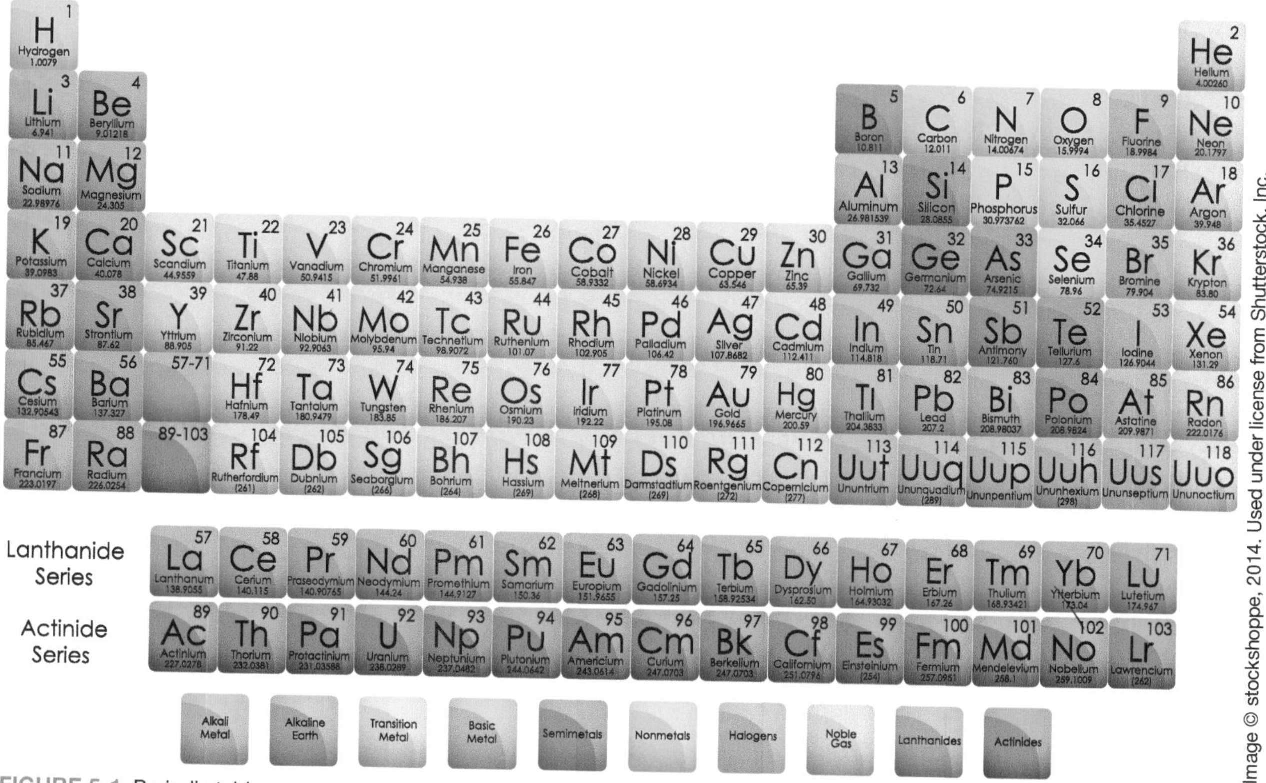

FIGURE 5-1 Periodic table

nonmetals), the bonds among them are called *covalent*. Two or more atoms of the same element can combine to form gaseous compounds like oxygen (two oxygen atoms), ozone (three oxygen atoms), hydrogen (two hydrogen atoms), and nitrogen (two nitrogen atoms). The bonds among these atoms are covalent.

A final note on the periodic table: Some of the elements, called *noble gases*, are in the far right column of the table. These elements, like helium (He) and argon (Ar), are so stable that they will not combine with other elements. They will not gain or lose or share electrons with other elements. They are inert.

Toxicology

Paracelsus, the so-called Father of Toxicology, studied poisons in the 1400s–1500s AD. Poisonings were one of the most popular methods of homicide at the time. To kill someone with poison did not require overpowering strength, a bloody crime scene, or even a face-to-face confrontation. Also, it was hard to trace the killer. Kings employed tasters to pretaste foods prior to royal meal consumption.

The science of toxicology deals with chemicals that might be poisonous.[2] Toxicologists recommend maximum doses and avoidance of exposures to very strong chemicals that might be poisonous in very small doses, and together with physiologists, they determine the health effects on life forms. Testing of chemicals is usually performed on lab mice where control groups can be compared with those exposed at various dosage levels. The results of such experiments are extrapolated to humans, and maximum contaminant levels are postulated. *Voluntary human testing* is also performed on potentially nonlethal chemicals. Also, data on disease and known exposures to chemicals (epidemiological studies) also add information on toxicity. Epidemiological studies, in contrast to exposure of lab mice, are based on data from past human exposure, but human exposure assumptions may be uncertain, lacking control groups and elimination of other risk factors.

When the analyses of chemicals of concern, the exposure assessment, and the toxicological assessment are completed, a risk assessment is performed to determine whether an environmental impact is realized. The impact must be tied to a source of toxin, a pathway that the toxin may travel from the source, and a receptor that is exposed to the toxin. Finally, the toxin must be in sufficient amount to cause a harmful affect. That amount would vary depending on the age, physical condition, immune system response, and other risk factors.

References

1. Scerri, E. R. 2011. *The Periodic Table: A Very Short Introduction*. Oxford: Oxford University Press.
2. Lu, F. C. 1996. *Basic Toxicology: Fundamentals, Target Organs, and Risk Assessment*. Washington, DC : Taylor & Francis.

6 Geology and the Environment

Geology is the study of the Earth, mostly the lithosphere, from the standpoint of Earth materials, processes, and history. Environmental science benefits from geologic studies by providing a framework for the environment in which we live, namely, the *structure of* the Earth, movement of crustal plates, material and energy resources, and Earth processes.[1]

Structure of the Earth

Our environment is influenced by the structure or framework of surface rocks and soils, the topography (landforms), and, to some extent, deeper lithospheric layers that influence volcanoes, plate tectonics, our magnetic field, and the movement of earthquake waves through the Earth's interior.[2]

Most pertinent to environmental science is the upper layer of the lithosphere, the crust. This includes the lighter crust of the continents, its cover of sedimentary rocks, and lava and volcanic gases that spew onto the surface through volcanic vents. The more-dense crust under the oceans is the site of mid-oceanic ridges exuding abundant geothermal energy. At great depths in oceanic crust are hydrothermal vents, the origin of a food chain that relies on chemo- rather than photosynthesis.

PLATE TECTONICS

During the late 1960s, a geologic theory was finally accepted that helped to explain the partly accepted theory of *continental drift*. Most geologists were puzzled by the "*jig-saw puzzle*" fit of continental margins, the correlation of geologic formations and fossils on adjacent continents that were now separated by oceans, and rocks that were clearly formed in climates that did not match their current latitudes (rocks from glacial deposits now in equatorial regions) **(Figure 6-1)**.[3]

Also in the 1960s, a second theory which also helped to explain continental drift was proposed: the theory of *sea-floor spreading*. Knowledge of mid-oceanic ridges and isotopic dating of rocks on either side of them indicated that new oceanic crust was being formed from recycled crust in the mantle. This new crust issued from volcanic vents at the underwater crests of the mid-ocean ridges. The age of the rocks away from the crests was progressively older, indicating that once formed the new crust was moving away from the crests—spreading the seafloor over time.

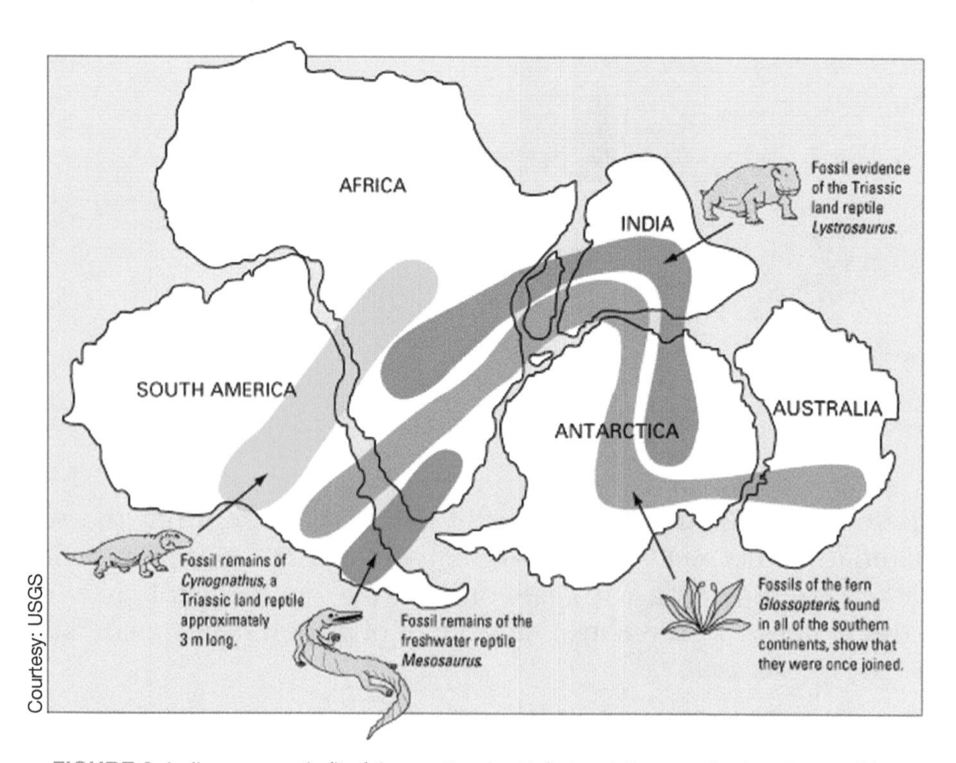

FIGURE 6-1 Jig-saw puzzle fit of the continents. Unfortunately, a mechanism that could explain the perceived movement of continents was not physically possible, namely the drifting of continental crust through oceanic crust. While a graduate student in 1963, the senior author was cautioned not to admit the acceptance of the theory of continental drift for fear of his professors' disapproval

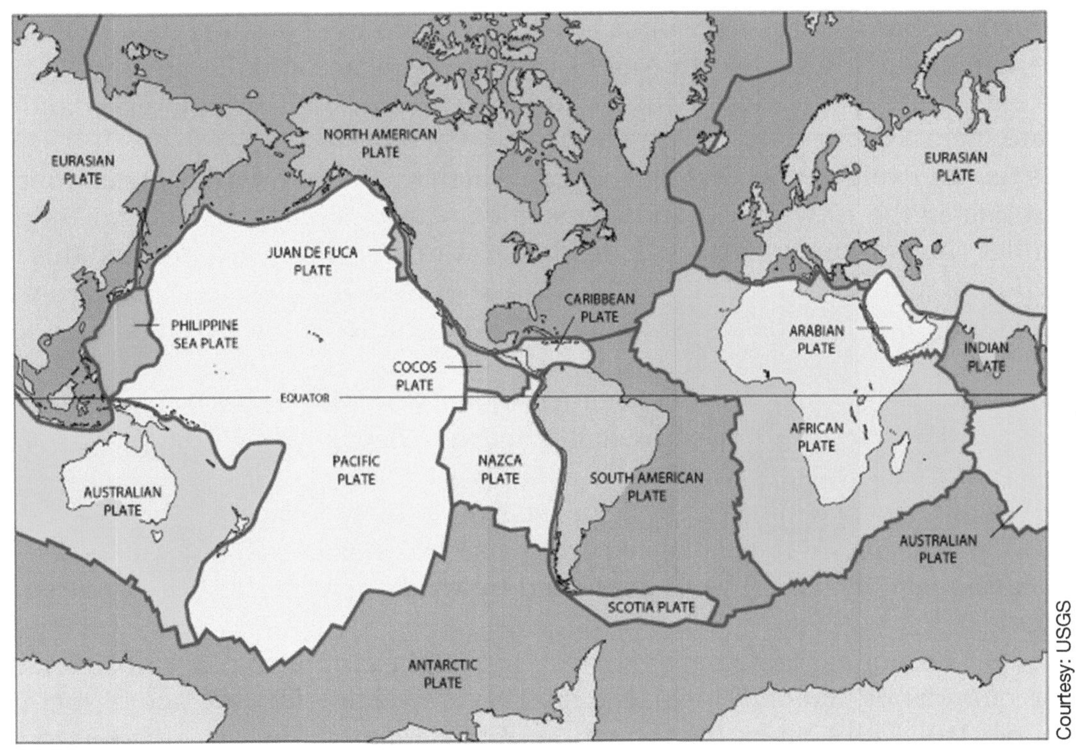

FIGURE 6-2 Tectonic plate boundaries. The new theory (of plate tectonics), as good theories should, explain many things in addition to the theory of continental drift. This model of plates also explained why very old continental crust was found (over 4 billion years old), but the oldest oceanic crust was only 350 million years. The answer was that older oceanic crust was subducted under other plates and destroyed. This theory also explained the location of earthquakes and volcanoes at plate boundaries

So a new theory to explain continental drift and sea-floor spreading was born, called *plate tectonics*. Instead of continental crust moving through oceanic crust, a new model of the crust was proposed. A group of crustal plates consisting of both continental and ocean crust moved. But, if new crust was being formed, where did the excess crust go? Additional oceanographic data identified zones where excess crust dove down below other plates and returned crustal material to the mantle. These zones were called *subduction zones*. The plates were analogous to conveyor belts that raised new crust at mid-oceanic ridges and returned old crust to the mantle at subduction zones (**Figure 6-2**).

The theory proposed three main kinds of plate interactions:

1. convergent (plates colliding head-on),
2. divergent (plates separating, as at mid-ocean ridges), and
3. transform (plates side-swiping each other laterally.

The *convergent plates* could be at subduction zones, where the edges of heavier oceanic crust dive below lighter continental crust (as along the Japanese Islands where the Pacific plate dives below the Asian plate). This produces deep-seated earthquakes and generates volcanic activity above the subsurface plane of a subduction zone.

Convergent boundaries could also be at collisions of two plates, both with edges composed of continental crust. In this case, the continental crust would buckle and form mountain ranges (as when the Indian subcontinent rammed into the Asian plate and formed the Himalayas).

Transform boundaries are like those along the California coastline, where the Pacific plate rubs laterally along the North American plate. This accounts for the line of active, earthquake-generating faults in California and the volcanic activity in Oregon and Washington State.

What was not explained at first was volcanic activity located within plates (not at the boundaries), as in the Hawaiian Islands. This is explained by the existence of "hot spots" in the mantle. The movement of the Pacific plate toward the northwest justified this model. This direction of movement explained why the age of the volcanoes in Hawaii from northwest to southeast was successively younger. Newer volcanoes were added southeast of the older ones.

Currently, the big island, called Hawaii (nicknamed the Big Island), has the youngest volcanoes and the only one that is presently active (Kilauea, at the southeastern end of the island chain).

What was also not explained at first was seismic activity located within plates (not at boundaries), producing the New Madrid earthquake of 1811–12 and the Charleston, South Carolina, earthquake of 1886. This is explained by weaknesses in the lower crust and the upper mantle.

What was also not explained at first were mountain ranges that are now located far from plate boundaries, like the Appalachians. This is explained by the age of the Appalachians (upper Paleozoic), when the formation of those mountains was adjacent to a plate boundary. It also explained the source sediment for the Appalachians as coming from the African plate that had been adjacent to the North American plate.

Consequently, plate tectonics does not explain everything, but it does justify continental drift and the locations of most earthquakes, volcanoes, the absence of very old oceanic crust, and geologically young mountain ranges.

Material and Energy Resources

Geologists are the finders of material and energy resources in the Earth's crust. They understand where to look for coal, oil and gas, ores of metals and uranium, construction materials (sand and gravel, clay), groundwater supplies, and sites having ideal foundation properties for buildings, dams, bridges, landfills, reservoirs, and power plants.[4,5]

The location of both material and energy resources is no secret. It would be nice if one were to take a lump of any nearby crustal rock or soil and get the resources we need. There are gold and other potentially useful materials in mostly every rock you might find in easy reach. However, the amount of each material is not economically extractable, unless it were already concentrated by Nature.

Geologists know the features on the surface and in the subsurface that indicate where, with a high probability, natural concentrations of useful materials are located **(Figure 6-3)**. Minerals can be concentrated in fractures, oil and gas can be concentrated where structural and stratigraphic traps are present,

FIGURE 6-3 Surface mining for metals

and abundant groundwater may be found in permeable beds below the water table. If one pans for gold in a stream and finds flecks of gold in the sediment **(Figure 6-4)**, there is good reason to prospect in bedrock areas upstream in search of the mother lode. *Sand* and *gravel* are found in stream channels and terraces and in beaches where running water or waves have sorted the sediments into uniformly-sized particles.

FIGURE 6-4 Panning for gold

Earth Hazards

Geologists can identify places that are prone to earthquakes, active faults, volcanoes, floods, subsidence from underground mines and fluid extraction, and permeable soils or rocks that may transmit leachate from waste-disposal sites. In studying and mapping crustal materials and observing Earth *processes*, signs of activity of our dynamic Earth alert us to hazards that can cause adverse health, accidents, property damage, and death. In some cases, human apathy and ignorance can make the natural hazards more dangerous. Human activities that make geologic hazards worse include construction:

FIGURE 6-5 Stromboli, an active volcano. Note town at lower right

- on a floodplain,
- at the foot of active volcanoes **(Figure 6-5)**,
- at the foot of landslide-prone areas,
- over abandoned mines,
- in areas prone to hurricanes, tornadoes, and tsunami, and
- astride active faults.

Choosing sites for landfills, septic systems, nuclear power plants, dams, and large buildings should take into consideration the natural properties of Earth materials and the probability of Earth hazards prior to construction. Not only will improper siting cost more money eventually, but the improperly chosen sites may increase the probability of environmental problems.

Earth Processes

The Earth is dynamic. As a result, gravity, surface and groundwater flow, tectonic plate movement, rock faulting, weathering and erosion, glaciation, and volcanic activity can cause changes in the environment that are detrimental to a safe and productive environment, and inhibit sustainable growth. A geologist can also identify deposits of soil and rock that are clues to the natural processes of the past, which can reoccur in the present and future. In the past one million years or longer, there has not been a geologic process that had not occurred previously.

REFERENCES

1. Keller, E. A. 2012. *Introduction to Environmental Geology.* Upper Saddle River, NJ: Prentice Hall.
2. Lowrie, W. 2007. *Fundamentals of Geophysics.* Cambridge: Cambridge University Press.
3. Frisch, W., M. Meschede, and R. Blakey. 2011. *Plate Tectonics: Continental Drift and Mountain Building.* Heidelberg, Germany: Springer.
4. Shackleton, W. G. 1986. *Economic and Applied Geology: An Introduction.* London N.H.: Croom Helm.
5. Riley, C. M. 1959. *Our Mineral Resources.* New York: Wiley.

7 Biology and the Environment

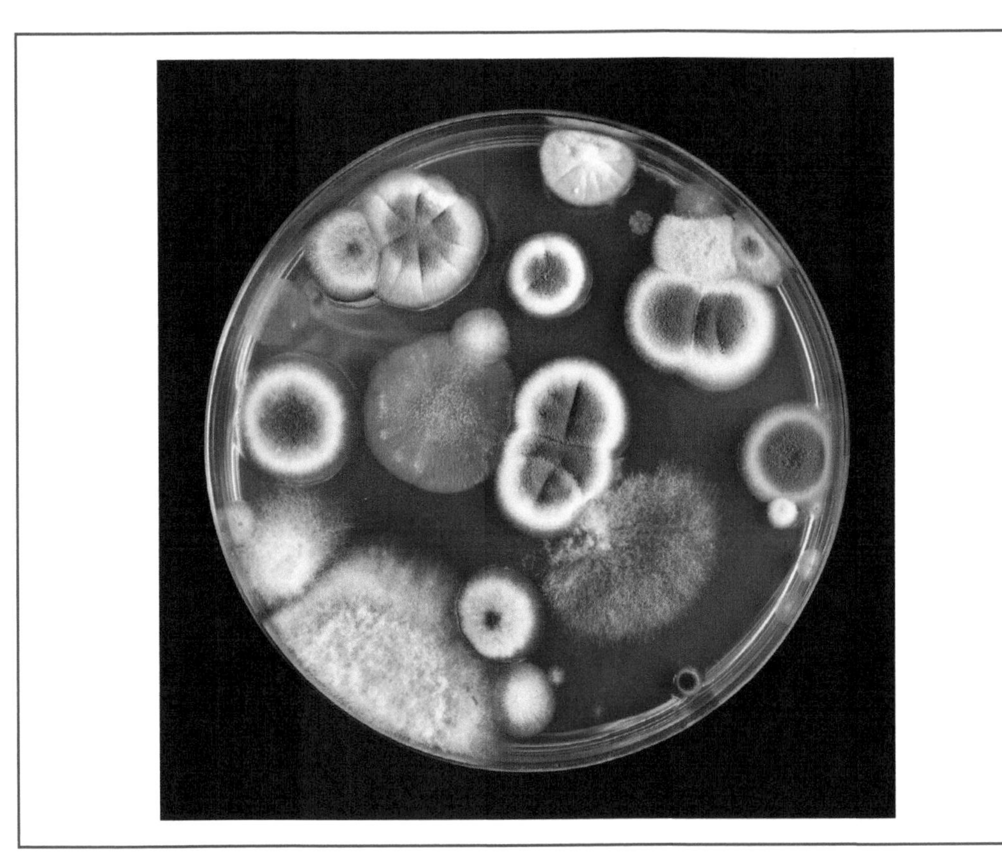

Photosynthesis (and Chemosynthesis)

The most important environmental benefit is the process known as *photosynthesis* **(Figure 7-1)**.[1] Photosynthesis is a free service of Nature. The Sun provides the energy, and green plants provide the mechanism to capture that energy and put it into useful edible form. Plants, as producers, manufacture glucose, an edible source of energy and a raw material for the manufacture of other food products (proteins and fats). A bonus by-product of photosynthesis is the formation of free oxygen (O_2), produced in the same chemical reaction that produces glucose. Organisms can use the oxygen, by the process of respiration,

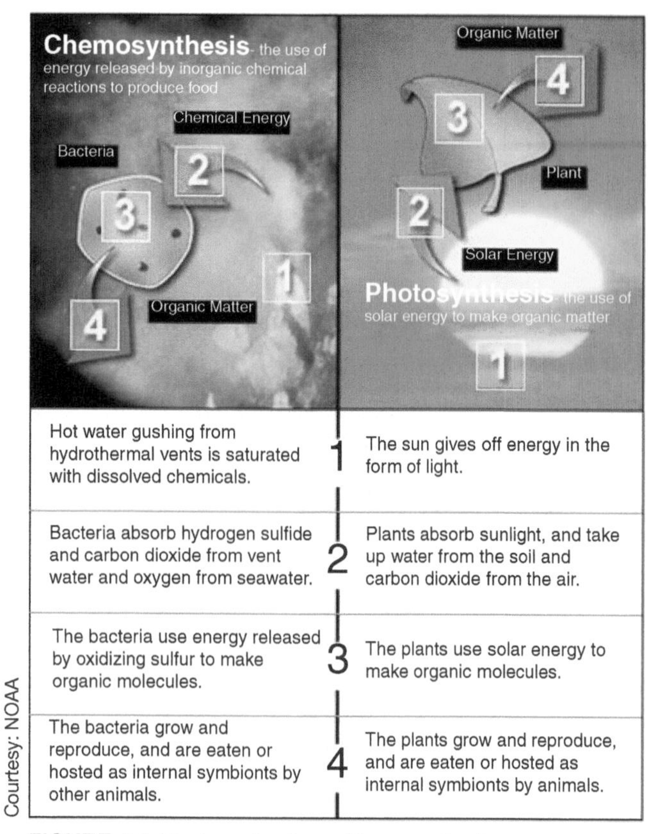

FIGURE 7-1 Photosynthesis vs. Chemosynthesis

to release the energy stored in foods to power and grow their bodies. Aside from sunlight, plants need only water and carbon dioxide to perform photosynthesis.

A similar process called *chemosynthesis*[2] **(Figure 7-1)** is used by producers in the deep ocean, beyond the depths that sunlight can penetrate. Here, at depths of more than 600 m, bacteria use gases from hydrothermal vents in the ocean bottom, which contain hydrogen sulfide and carbon dioxide. The bacteria break the bonds in the hydrogen sulfide molecules to produce carbohydrates needed by consumers in the deep ocean *habitats*.

The production of glucose and oxygen by plants for animals by photosynthesis (and in part by chemosynthesis) is recycled by animals through respiration to produce water and carbon dioxide, ready for use in photosynthesis again. This recycling aids, with a few exceptions, all life on earth. It is a remarkable environmental service.

Ecology

Ecology is a subdiscipline of biology that deals with the relationships of life forms to their environment. This part of biology is most important in environmental science. It could also be called environmental biology or the study of ecosystems.

Habitats and Niches

A habitat is where an organism lives. Deer live in forests and grasslands. Earthworms live in the soil. A leopard lives on a savanna. A hawk or eagle lives high over areas of potential prey in an aerie. A shark lives in the ocean. A frog lives in a freshwater pond. A human city-dweller lives in an urban area. A cactus plant lives in a desert. These are habitats, i.e., areas where the organism has space to live and reproduce, food and water to eat and drink, protection from predators, and shelter from adverse environmental conditions.[3]

A niche, on the other hand, is the role the organism plays in its environment. It is its lifestyle. Although the habitat of honey bees comprise the beehive and an area around the hive where pollen and nectar can be gathered from nearby flowers, the honeybee has several niches depending on its role in the hive.

Female worker bees do all the work, except mating. They build the nesting cells, the comb, where honey is stored and where the queen lays her eggs, they gather pollen and nectar, make the honey, take care of the queen **(Figure 7-2)**, tend to the young males or drones, and the larvae, and defend the colony. If the colony is threatened they swarm with the queen to establish a new beehive.

FIGURE 7-2 Bees in beehive with queen in the middle

The sole queen bee in the hive makes sure that she is the only fertile female by emitting pheromones to keep all other worker females sterile. After she emerges at birth from a specially built cell in the hive (giving her the niche of the queen), she checks all the other cells and kills any other potential queen rival or she must fight to the death any queen that had already emerged. Queens can sting many times, but the female workers can sting only once and die as their barbed stingers cannot be withdrawn. They must have their stingers intact to remain alive.

Oh yes . . . the male drone. This bee's niche is only to mate with the queen. The drones' existence does not involve work around the hive, collecting pollen or nectar, tending to the others, or defending the hive. After mating with the queen, the females prevent the drones from entering the hive—starving them to death.

Ecosystems

Ecosystems are ecological systems comprising communities of organisms in a given area interacting with themselves and their physical environment. The study of ecosystems is called ecology.[3]

Organism Interactions

To maintain the most efficient system, a form of ecologic capitalism takes place. As with economic capitalism, the dominant theme is competition. Darwin called this "*survival of the fittest*." In ecosystems the fittest are those who can survive (especially through reproductive years), use resources wisely, and most importantly adapt to the existing and inevitably changing environment.

Organism competition can take place among members of the same species (intraspecific), and among members of different species (interspecific). *Intraspecific competition* in most cases is limited to competition without killing members of the same species, although in some social orders this does take place (cannibalism). *Interspecific competition*, killing and eating members of other species (predation) is common among carnivores for survival.

Interspecific competition without predation can take the form of better access to food resources. We are familiar with the proverb "The early bird gets the worm." So birds that wake up and start gathering food earlier than others may have the competitive edge on food gathering. Another competitive arrangement also involves studies by Darwin involving the evolution of bird beaks. Different beaks allow certain bird species to open different seed shells more easily. Also, nocturnal animals (active during the night) can gather food from the same food source as other animals that are active during the day without competing face to face.

Ecosystems provide other resources, other than animal food. Nesting places, protection from predators, soil nutrients for plants, climates for survival (with or without seasonal migrations), are some of these resources. All of these interactions are beneficial to the sustainability of the ecosystem when they help to maintain optimum population sizes, a continuing supply of resources, and adaptations to potential changes in the environment.

Predator/Prey Relationships

A predator is an animal that kills and eats another animal, its prey **(Figure 7-3)**. Both interact. The predator, without the prey, would die of starvation. The prey, without the predator, may multiply so rapidly that its food source might be used up, die out, and the predator in turn would die without its food source. So a balance in Nature must be maintained for a sustainable relationship among interacting predators and prey.

An equilibrium must be established. This equilibrium, however, is dynamic, as on Earth change is the rule and not the exception. The dynamics involve evolutionary changes that favor maximum efficiencies. Predator skills involve speed to catch its prey, keen senses to hunt its prey, camouflage to lie in wait, sharp teeth, claws, and strength to subdue its prey, etc. Prey skills to avoid capture might also include speed to escape, keen senses to prepare to escape, camouflage to avoid detection, and defensive means to repel its attacker. The survival of predators and prey would be favored by those predators that were great hunters and those prey who were skilled at escaping capture. Therefore, the next generation of predators and prey would have improved skills.

FIGURE 7-3 Lion (predator) attacking water buffalo (prey)

FIGURE 7-4 Sea horse resembling seaweed

Camouflage or *mimicry* can help prey to hide from predators, like the stick insect that looks like an unappetizing twig, the butterfly whose coloration looks like the terrifying eyes of a predator, or the sea horse that resembles seaweed **(Figure 7-4)**.

Plant Competition and Protection

Although predator–prey interactions refer to animals, the relationship between animals and the plants they may eat is analogous. For example, trees that protect their leaves by growing taller than most grazing animals can encourage increased populations of giraffe, whose long necks enable them to browse on leaves high up in the trees. Poisonous leaves or roots can protect them from being completely eaten up.

Plants can compete with each other for space, light, and water. This is most noticeable in gardens and lawns, where humans grow plant species that are alien to the area. In gardens and lawns (and croplands), native species (we call weeds) will try to regain a foothold on the land. Plants can spread to exploit space to grow and spread, absorb sunshine, and capture mineral resources and available moisture. Some trees, like the strangler fig, can wrap their entangling branches around other trees for support and access to sunlight above the forest floor.

Symbiosis

Some creatures like to live together for mutual benefit or selective harm. The coral animal builds a reef of limestone in shallow tropical seas. The coral polyp can catch sea creatures with stinging tentacles, but it needs additional nutrition provided by colonies of one-celled plants, algae. The alga finds a home in the reef and, through photosynthesis, can produce sugars to supplement the diet of the coral. The coral, therefore, must build upwards to stay in shallow water for sunlight to penetrate to the algae. Spanish "moss" (not a moss, but a flowering plant) drapes its leaves on tree branches, which shares available sunlight and protects the tree from wind damage during storms. *Symbiosis*, an arrangement where both host and guest benefit, is called *mutualism*.

Other organisms move into the bodies or shelters of other organisms to reap benefits to themselves but harm to their hosts. This form of symbiosis is called *parasitism*. Diseases can be caused by parasitic pathogens that live with the host and slowly feed on their

FIGURE 7-5 Venus flytrap

host's body or secrete waste fluids that poison their host. This process is self-destructive as a parasite that kills its host must die or quickly move to another host. Moving to other hosts to spread disease is the process we call contagion, or more widely spread an outbreak, or even more widely spread an epidemic, or with a worldwide spread a pandemic.

Harmless or nonbeneficial symbiosis is called *commensalism.*

Attraction

Intraspecies (within a species) or interspecies (between species) attraction can have positive effects.

- Animals can attract mates by coloration, mating rituals and calls, and body odors.
- The *Venus flytrap* **(Figure 7-5)** and the pitcher plant can attract insects in their traps to add micronutrients to their diet. These plants are not considered animals as their main diet is still derived from photosynthesis.
- *Flowering plants* attract insects to help disperse and deliver pollen to another flowering plant of the same species. They attract the "unsuspecting" pollinators with aromas and nectar, and some mimic the shape and coloration of female insects.
- *Fruit (and nut) trees* can attract animals to eat the fruits (containing seeds) and deposit the seeds and their droppings (as fertilizer) at a distance from the tree. As the tree is stationary, this process helps to extend its geographic range.

Energy Flow Through Ecosystems

We owe most of our energy for life on Earth to the Sun. The Sun is a rather mediocre star of hot gasses in a galaxy called the Milky Way. The great gravitational pull of the sun on its gasses provides the energy to fuse hydrogen atoms and form helium, with the release of great quantities of energy (by this atomic fusion).[4] Radiant energy from the fusion is emitted into space at a speed of 186,000 miles per second (~300,000 km/second) and reaches the Earth, about 94 million miles (~150 million km) away, in about 8 minutes. Little energy is lost between the Sun and the Earth in the vacuum of intervening empty space, but when it hits the Earth's atmosphere much of the energy is absorbed as heat. Only about 1% of the energy reaches the plants, and the plants use only some of the red and blue rays of the visible spectrum to perform photosynthesis.

To perform photosynthesis, plants must be green in color. Why are plants green? Plants are green to reflect green light so that they can more efficiently absorb the red and blue light for photosynthesis. The green color is from the chemical chlorophyll, a photoreceptor. Chlorophyll, oriented properly by a protein molecule, allows the plants to absorb light efficiently and, thereby, convert water and carbon dioxide into oxygen and carbohydrates (glucose). If it were not for photosynthesis, life forms (except for some bacteria) could not exist without oxygen to breathe—either underwater (dissolved oxygen) or on land. Plants also use oxygen when there is no sunlight. Also, if it were not for photosynthesis, life forms (except for some creatures in the deep, dark ocean) could not exist without the formation of carbohydrates. Nonplants either eat carbohydrates directly or eat food materials (proteins, fats, fiber) that are synthesized by other plants and animals. The food contains energy for life in the chemical bonds originally created by photosynthesis. The carbohydrates and other synthesized food materials also provide all organisms with body materials for growth and repair.

Although the end products of photosynthesis are essential for life, trace amounts of mineral matter are also needed for more complex carbohydrates, proteins, fats, and skeletal materials for organisms. The source of many of these trace minerals are in the soil or dissolved in water. Thus, photosynthesis does not explain all aspects of the ecosystem—the nonliving (inorganic) part of the ecosystem plays a key role. The water, carbon dioxide, and trace metals are the source materials of the ecosystem –and will be the end products as well.

Plants are producers of food (carbohydrates) at the lowest tier of an energy pyramid (**Figure 7-6**). These tiers are called *trophic levels.*[5] Continuing up the pyramid to higher trophic levels are the animals. Instead of producers like plants, animals are consumers (they eat!). The next trophic level comprises the herbivores, or animals that can only eat plants for nutrition (e.g., cows, deer, sheep) Next is the third trophic level, comprising carnivores (e.g., foxes, wolves, snakes), which can eat only animals for nutrition, and omnivores (e.g., raccoons, humans), which eat animals and plants. Some ecologists characterize the top, large carnivores (e.g., lions, tigers, raptors, such as hawks and eagles) into a separate trophic level. These top carnivores eat carnivores, herbivores, and omnivores.

With each rise in trophic level, energy is transferred from the lower to the next higher level. The consumer ingests the food materials that contain energy. The consumer uses some and stores some. If the consumer is subsequently eaten, the stored energy in its body is passed up the food pyramid to the next higher trophic level. Unfortunately, 90% of the energy in each transformation is lost (to heat). Consequently, more biomass must be eaten with each higher tier of the food pyramid. As the food is eaten and the energy passes from one trophic level to the next higher level, we call the *chain* of passage a

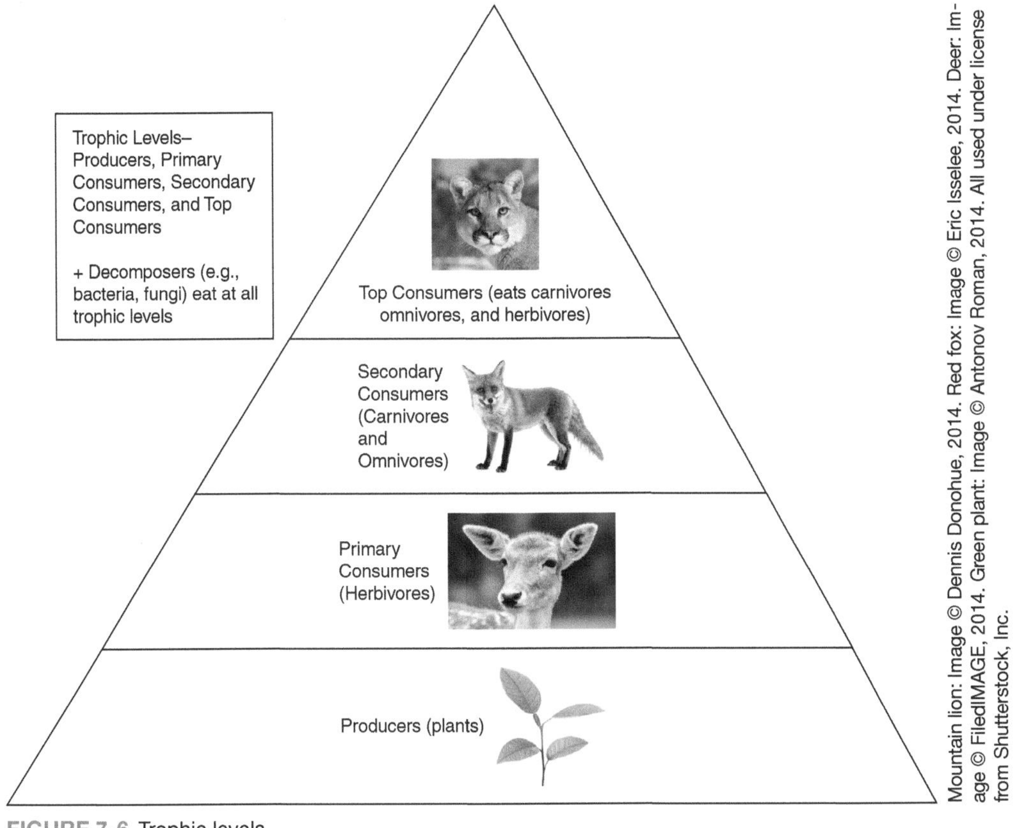

FIGURE 7-6 Trophic levels

food chain. Obviously, different organisms have different diets, so food chains vary. When multiple food chains overlap, we call these *food webs*.

Humans can eat both animals and plants, though some choose not to eat animals (vegetarians). Some humans, called *vegans*, will not eat any products from animals (e.g., milk, eggs). Some humans will eat only vegetables that are fruits, which fall from the tree or vine, because the harvesting of these vegetables does not kill the plants. Such groups of people exclude certain foods for several reasons. They maintain that it is inhumane (unethical) to kill animals, which have a right to live and would feel pain when killed. Another reason to eat lower on the food pyramid is to conserve energy, as less energy is lost with fewer trophic level transformations. Not only is energy conserved when consumers eat at the lower trophic levels, but less biomass is required to feed large populations of people.

After the food is produced and consumed, the energy in the foods is used for the life processes of all the producers and consumers. Energy flows through and up the food pyramid in the ecosystems. What happens to the energy? It is not destroyed according to the scientific law of conservation of energy; instead—it is eventually transformed into heat energy that cannot be used again effectively (entropy).

Matter in Ecosystems is Recycled

Leftover foods (with leftover residual energy) would be a problem if it were not for the detritus feeders and decomposer organisms (scavengers). *Detritus eaters*[6] eat the leftover organic materials and break them down for further processing by decomposers. The decomposers, including bacteria and fungi, take the organic compound leftovers and reduce the compounds to water, carbon dioxide, nitrogen gas, and metallic micronutrients. They effectively return many of the "waste" products to the soil, atmosphere, and water. The waste products in their simple form can be used again by plants to restart the cycle. Almost nothing goes to waste.

Recycling of elements can be traced through the Earth spheres. Carbon can be recycled in the following way: carbon dioxide, to plant tissue, to animal flesh, to bacterial breakdown products, to calcium carbonate, to limestone, to weathered limestone soil, and back again to carbon dioxide. Water can be recycled in the following way: rain, to lakes, to ingestion by fish, to ingestion by water-fowl, to human hunters, to urine, to sewage treatment plants, to streams, to water vapor through evaporation, to clouds, and to precipitation again. Nitrogen can be recycled in the following way: air, to pore spaces in the soil, with nitrogen fixing bacteria to nitrates, to plant uptake (grasses), to grazing sheep, to sheep feces, to bacterial decomposers (denitrifying), to nitrogen gas back into the atmosphere. Phosphorus can be recycled in the following way: minerals in rock could be released by erosion, and enter the soil, absorbed by plants, eaten by animals, and their bones fossilized back into rock.

With every breath, we inhale air that is 79% nitrogen gas. We use about 21% of the air for respiration (from the oxygen) and less than 1% for triggering our breathing reflex (from the carbon dioxide). We use no nitrogen, which is exhaled along with carbon dioxide and water vapor. All that the nitrogen does is to help exercise our lungs. Yet we need nitrogen in compounds including fats, proteins, and genetic materials. Plants also cannot use nitrogen gas directly. So, where does this nitrogen come from? It comes from nitrogen fixing bacteria[7] in the soil associated with certain plants (legumes). At the roots of legumes, nitrogen-fixing bacteria take the nitrogen from the soil gas and convert it into nitrogen compounds that can be used by plants and animals. Therefore, nitrogen fixation is an environmental service provided by certain bacteria and the leguminous plants that provide the bacteria's habitat.

To recycle the nitrogen back into gas molecules, denitrifying bacteria reduce the nitrogen compounds back into components to be used again for nitrification.

Earth's Biomes

In different areas with varied climates, soils, and other physical influences, the ecosystems are quite different from one another. As noted previously, all ecosystems having characteristic climates, soils, and other physical influences can be included into distinct geographic areas called biomes.[8] Biomes can be terrestrial (land based) or aquatic (water based). They may be microscopic or macroscopic in size. New aquatic biomes are being discovered in the subsurface of ocean basins, where microscopic organisms thrive without light and endure temperatures exceeding 700°C. Not all biomes are so unusual; fairly common examples include desert, grassland, wetland, forest, and stream. Biomes are classified based on temperature and/or availability of moisture, and in some cases based on their soils (which are influenced themselves by temperature and moisture).

Deserts

A biome that has low annual rainfall (<10 inches, 25 cm) is called a *desert*.[9] Deserts need not be hot; the Gobi desert of northern China is northwest of Beijing at about the latitude of Seattle, Washington, and lies about 3,000 to 5,000 feet above sea level. Average annual temperatures are around freezing (25–35°F, 2.5°C), with freezing temperatures and snow and frost common in the winter. Most deserts, however, are warm, although cold nights on high-elevation desert plateaus (southern Arizona and Nevada) are common. Wind is common, which in places is characterized by dune sand. Cloud cover is uncommon. The annual fluctuation of temperature is characterized by higher temperatures in the summer and lower temperatures in winter. Precipitation is almost nonexistent in summer, and low in winter.

Although deserts have low accumulations of precipitation in any one year, when it does rain, these biomes may experience brief but torrential showers. The rainfall can be so rapid and intense, that little of the water infiltrates into the ground. Instead it runs off quickly into otherwise dry gulches that can be quickly transformed into deadly washes of turbulent (nonlaminar) and turbid (full of sediment) waters.

Why do many deserts seem to be located around 20° north and south latitudes and/or downwind from mountain ranges? Droughts are influenced by falling air currents at these latitudes allowing the atmosphere to warm and retain moisture. Dry climates are also influenced by the rain shadow effect. In traversing the mountain ranges, air must rise up to go over them, where it cools, and thereby loses available moisture.

As this climate is relatively harsh to many plants and animals, certain species that are well adapted to the desert habitats thrive without competition from, say, species adapted

© Alan Jacobs

FIGURE 7-7 Desert flora near Phoenix, Arizona

to habitats with wide moisture tolerances (**Figure 7-7**). A *cactus* plant can outcompete an oak tree and a camel can outcompete a hippopotamus in the desert. Nevertheless, these specialized habitats can support low numbers of each species.

The resulting droughts favor plants like cacti and noncactus succulents (no leaves), which reduce their propensity to transpire and lose moisture (evaporate on leaf surfaces). When the spring rains come, the flowering period is brief and intense, and insects (pollinators) proliferate. Burrowing and/or nocturnal *animals* (e.g., lizards, snakes, rodents, burrowing owls) are common in deserts. They take advantage of the cooler temperatures at night and protection in their burrows in the treeless terrain.

Ectothermic animals (e.g., reptiles), which must regulate body temperatures from their environments, must adapt to the dramatic temperature fluctuations of the desert. Reptiles can warm their bodies in the Sun on rocky ledges in preparation for the cool nights, when nocturnal prey are active above ground. Endothermic animals (e.g., mammals and birds), which regulate body temperatures internally by perspiring (a cooling process), shivering, or having body hair, must also adapt. They stay hidden in burrows during the day and venture to the surface at night to hunt and forage.

Ground water is present in the desert at depth, or in shallower areas at oases. Surface water evaporates easily.

Humans who inhabit developments in the desert adapt by pumping water from subsurface aquifers, "borrowing" water from lakes behind dams, and diverting rivers from mountain sources. Las Vegas borrows water from Lake Mead, behind Hoover Dam, and must clean up wastewater and replace it each year. Some of the water from the Colorado River has been diverted into Southern California, where it is used to recharge an aquifer supplying water to Los Angeles. Because of the climatic scarcity of water supplies in the desert, many desert communities restrict frivolous water use (lawns and washing cars). For agriculture, traditional surface irrigation (in furrows) has been replaced by spray and drip irrigation, delivering water more directly to plant roots.

Grassland (Prairie or Steppe) and Savanna

Historically, the central United States was comprised of grassland, tall grass prairie towards the east and short grass prairie west to the Rocky Mountains. The tall grass was so tall that a rider on horseback could take the ends of shoots of grass on one side of the horse and tie it in a knot over the saddle with the ends of shoots of grass on the other side of the horse. Almost all of the grasslands, especially the tall grass prairie, had been converted to croplands during the 1800s. Western prairies became grazing lands. Soil abuse and erosion, however, produced "the Great Dust Bowl of the 1930s".[10] Better soil conservation techniques returned abused grasslands to productivity. Natural grasslands (**Figure 7-8**) are rare now and are being protected as refuges for grazing herds of wild deer, bison, elk, and other ruminants.

FIGURE 7-8 Grassland

Why were grasslands (tall grass prairie) so productive when they were converted to agricultural lands? Annual grasses added organic matter to the soil, fine-grained wind-blown silt (loess) made a good loam (soil of mixed-sized particles) that held moisture and nutrients, and natural fires kept the woody trees and shrubs from taking over the grassy areas. Short grass prairies were used as grazing lands. When the lands were overly grazed, invasive species of shrubs (e.g., sagebrush) reduced the biomass of edible grasses, which support domestic and wild *herds* of ruminants.

Grasslands receive 25 to 75 cm (10–30 in) of annual rainfall. Most of the rain and snow fall in the winter and early spring. Temperatures are cold in the winter and hot in the summer. Fires would be common on natural grasslands, but in areas of human habitation and agriculture, fires are controlled. Compared with forested land, rainfall here is lower. Consequently, trees are rare and are restricted to the flanks of drainage courses.

The *savanna*,[11] a type of tropical grassland in near-equatorial parts of South America and Africa, and in northern Australia, has more trees (in spotty patches) and a higher (two times) annual rainfall of 50 to 150 cm (20–60 in) than grasslands. Temperatures are tropical (hot), and the monsoon rainy season is quite humid. Fires are common. The famous *Serengeti Plain* is a savanna in Africa, known for wild animal parks of grazing herds and their feline predators. Agriculture and domestic grazing animals (goats and sheep) are common.

In all grasslands, plants and animal species are diverse and great in numbers. The term *herds* indicates that each species of grazing animals has large populations. The rich soil supports burrowing animals. Birds of prey are abundant. They soar with the aid of up-drafts of air while looking for prey. They have unobstructed (treeless) views of the ground from aloft.

Mediterranean Shrublands (Chaparral)

When a climatic regime similar to savannas is present near the sea –as along the coastal areas of the Mediterranean, southern California, and areas in South America, Africa, and Australia—woody shrubs crowd out the grasses and form a chaparral.[12] This biome has

FIGURE 7-9 Chaparral

hot, dry summers, and cool, wet winters. Also, unlike the general flat topography of grasslands, the chaparral landscape is hilly and rocky, allowing well-drained soils and some shade.

Because of the dominance of shrubs, wildlife can take refuge on the surface or in the subsurface among stands of shrubs, harboring quail and rabbits (**Figure 7-9**). Shrubs and short trees have needle-like leaves, some waxy, to withstand periodic droughts.

The region of shrublands is well inhabited by humans. Historically, the vegetation encouraged the culinary arts. The soil and climate are optimal for cultivation of olive trees (olive oil and olives), aromatic herbs (basil, oregano, rosemary, thyme, sage), and vineyards. Grasses can support small groups of ruminants (not herds).

Forests

Forests, as one would expect, are dominated by trees. Forests, as one might not expect, have soils that are not suitable for crops. Most of the nutrient storage in forests is in the vegetation itself, not in the soil. Therefore, clear-cutting of forests for agricultural land yields poor crops after the sparse nutrients in the soil are quickly used up. In transforming forests to cropland by burning the trees and plowing them into the soil provides additional nutrients for only a few growing seasons.[13]

There is a great variety of forests depending on the climate (temperature and rainfall). There are tropical (hot), temperate, and boreal forests or taiga (cold northern/high altitude), both dry and rainy (rainforests). Species diversity varies from low, in the taiga, to very high in tropical rainforests.

Boreal Forests (Taiga)

Boreal forests are in higher latitudes or higher altitudes, which are colder.[14] Here, conifers abound, as the frost-resistant needles, especially those that stay green year round, can be productive through the cold winters. The conical shape of the tree and its flexible branches can support or shed the weighty snow and be resistant to high winds, without its limbs easily breaking off. Pollination of conifers is by wind transport, allowing seeds to form in protective cones without reliance on insects.

The boreal forests replaced areas of temperate forests during the last ice age. As the ice fronts receded and the temperatures rose, most boreal forests were replaced by temperate forests. Pockets of boreal forests still remain as *refugia* of the last ice age. There are four such areas in Ohio where one can see flora similar to that of areas hundreds of miles north (in Minnesota, Michigan, Wisconsin, and southern Canada). Tree species like the *tamarack* and bog plants like sphagnum moss still survive. The tamarack is a deciduous conifer whose needles turn yellow in the fall and fall off the branches **(Figure 7-10)**; green needles grow again in the spring.

© Alan Jacobs

FIGURE 7-10 Kent Bog refugium, South of Kent, Ohio

Temperate Forests

Temperate forests have milder winters and greater species diversity.[15] These forests are dominated mostly by deciduous trees. Several environments are found in these forests. On the forest floor, leaf litter and decaying tree bark and branches constantly accumulate. Here seedlings and wildflowers can start to grow and get sunlight in the early spring when trees are not in leaf. Here, too, decomposer organisms can breakdown the fallen plant debris and release nutrients for new growth. Another forest environment is the canopy. It is the highest level of the forest and is usually uniform in height, allowing sunlight to hit the tops of all adult trees. Nesting and roosting birds abound. The temperate forests are homes to many animal species (e.g., foxes, deer, raccoons, squirrels). Logging is common, which, when managed properly, can be sustainable with little damage to wildlife habitats. Clear-cutting and slash-and-burn practices destroy the ability of the forest to be self-sustaining.

Tropical Forests

Tropical forests (both dry and wet) have mild or no winters at all.[16] Without frost, plants can be green year round **(Figure 7-11)**. The tropical rainforest biome has the greatest species diversity. Flowering plants, birds, and insects abound. The arboreal environment is also diverse, with populations of primates (monkeys, apes, birds, and other animals) **(Figure 7-12)**. Rainforest plants have been harvested for the manufacture of pharmacological products to treat cancer, malaria, heart disease, bronchitis, hypertension, and a host of other diseases.

© Alan Jacobs

FIGURE 7-11 Tropical rain forest, Singapore Botanical Gardens

© Alan Jacobs

FIGURE 7-12 Young orangutan, Singapore Zoo

Tundra

In extremely high latitudes and altitudes, we find a wintry counterpart to the deserts of lower latitudes.[17] This biome is called *tundra*. Here the temperatures and precipitation amounts are extremely low. Snowfall, surprisingly, is minimal. This biome has been labeled by some a *"frozen wasteland."* Some of its soils are frozen solid year round

© Alan Jacobs

FIGURE 7-13 Model of an Iñupiat ice cellar, Iñupiat Heritage Center museum, Barrow, Alaska

(permafrost) and, if thawed, form waterlogged soils and shallow ponds. Sparse vegetation grows close to the ground. The wind is fierce. The growing season is short. High latitudes exhibit short day-lit hours in the winter (some weeks without any sunlight) and long day lit hours in summer (some weeks without nightfall).

The northernmost city of the United States, *Barrow*, is situated on the North Slope of Alaska, on the shores of the Arctic Ocean. Barrow is only about 1,300 miles (2,100 km) from the North Pole. Native Americans in this area (Iñupiats) fish and hunt whales. They take advantage of the permafrost by storing their meat and fish underground during the summer in hollowed-out pits (ice cellars) in the frozen soil **(Figure 7-13)**.

Plants include *dwarf willow* (growing close to the ground) and some birches. Otherwise, plants include grasses, sedges, and mosses, with a transition organism between a plant and a fungus called a *lichen*. Lichens are numerous in the tundra and help to break down rock into soil.

The tundra does not support a diverse group of animal species (lack of species diversity) and each species is few in numbers, with the exception of herds of caribou and, in the southern hemisphere polar regions, flocks of penguins. Large animals like bear (polar and grizzly) and small animals like rodents are present. Pesky insects are numerous in the short summer (50 to 60 days) and proliferate to pollinate flowers and decompose animal and plant wastes. The insects are food for migratory birds.

Wetlands

Transitional from terrestrial (land) to aquatic (water) ecosystems is the biome called a *wetlands*.[18] Wetlands are present on Earth, having various water salinities: freshwater wetlands (far from the sea) and salt- or brackish-water wetlands (at coastlines) called *salt marshes*. A salt marsh in the Gulf Coastal area of the United States is called a *bayou*.

The term wetlands now has a positive connotation because of the environmental services it provides—flood control, cleansing of contaminated waters, water storage during droughts, and habitats for diverse communities of wildlife. Previously, it had a negative connotation by those wanting to develop a well-drained tract of land for real estate development. Today, remediation facilities called *constructed wetlands* are engineered and built to clean up contaminated water. Some of these facilities are adjacent to facilities whose outflows are decontaminated after flowing through these artificial wetlands.

An example of a constructed wetland is a maze of connected channels adjacent to the Mahoning Landfill in Springfield Township, Mahoning County, Ohio (southeast of Youngstown). This wetland was constructed to eliminate ammonia from landfill leachate.

A second example is in a constructed wildlife refuge adjacent to a sewage treatment plant west of Delray Beach, Palm Beach County, Florida. After sewage treatment, the treated water is further purified by passing it through the wildlife refuge, where tropical birds, reptiles, and other wildlife find habitats **(Figure 7-14)**. Wildlife is so rich and diverse, that the site attracts human visitors, who can traverse the wetland on raised boardwalks with their cameras, binoculars, and curiosity.

Another transitional biome (between terrestrial and aquatic) is our national treasure, the *Everglades*. This landform is a transition between a wetland and a grassland. It is, in effect, a river of grass, as the water has a steady flow.

© Alan Jacobs

FIGURE 7-14 Great white heron in wildlife refuge Delray Beach, Florida

The Everglades

The largest wetland in the United States is located on the southern third of the *Florida peninsula*. It is called the Everglades. It is a wetland that extends from Lake Okeechobee at the north to Florida Bay at the southern tip of the peninsula. It has abundant wildlife, including alligators **(Figure 7-15 and 7-16)**. Human development that involves disruption of this extensive and valuable biome remains a threat to its sustainability. Sustaining this biome protects its habitats for wildlife, its ability to store water in times of floods and droughts, its ability to cleanse water flowing through, and as a safe harbor for migrating birds. How did this unique and extensive wetland form?[19,20]

The Florida peninsula, a land platform, rose from the sea tectonically (upward crustal movement) about 25 million years ago (mid-Tertiary Period). The peninsula during the remainder of the Tertiary was a time of marine deposition and erosion and occupation of the land by plants and animals.

After the Tertiary was the *Quaternary Period*. It was marked by the ice ages of the Pleistocene Epoch. The peninsula was submerged in part during interglacial times when polar continental ice sheets and mountain-valley glaciers melted and raised sea level (higher than it is today). During the interglacials the land area of Florida decreased, first making the peninsula narrower and less extensive to the south and then completely submerging

© Alan Jacobs

© Alan Jacobs

FIGURES 7-15 and 7-16 Alligators, Everglades National Park. Alligator Alley it below sea level. During glacial periods, conversely, the glaciers increased in volume and lowered sea level. This increased the land area of what is now Florida to make the peninsula wider and more extensive to the south than it is today. During the late Quaternary, humans began to occupy the land. Two thousand years ago the Glades people, followed by the nations of the Seminole and Creeks (Muscogee) populated the peninsula. After European settlers came, human occupation started to significantly alter the movement of water on the peninsula to suit their needs.

Much of the surface of Florida is characterized by flat terrain dominated by calcium carbonate-rich (lime) rocks and soil. The low elevations, nearness to the sea and relatively high rainfall results in a very high water table and numerous wetlands. A large inland freshwater lake (*Lake Okeechobee*) is present in the southern third of the peninsula, which drains to the southern tip of the peninsula into Florida Bay. The flow path of this giant wetland is sometimes referred to as a "river of grass." The official name is the Everglades.

If the ice ages return (without further tectonic uplift or depression) the land area would increase again. Land development would extend seaward but current seafront property would be farther inland. On the other hand, if the current interglacial results in more glacial melting, seaside cities and beach resorts would become flooded. Consequently, our future human habitation of this peninsula depends in part on sea level fluctuations. Since the current highest point in Florida is only about 350 feet above sea level, it would not take much global warming and glacial melting to almost submerge (except for extreme northern Florida) the entire state of Florida.

In the twentieth century, there was an increase in development of communities on the peninsula spurred by the lure of mild winter climates. Northerners wanted a place to escape from the cold and snowy winters. At first, the land was difficult to tame, with jungle-like vegetation, swampy foundations, alligators, snakes, biting insects, and wild cats. Development started with the establishment of farms, ranches, and citrus groves (orchards) because of long growing seasons, high rainfall, and almost absence of hard frosts.

Foundations for houses, roads, and rail lines were impeded by the presence of a high water table resulting in standing water in excavations. Consequently, extensive dewatering (draining) had to be done to make the land attractive for prospective human inhabitants. This altered the hydrology and adversely affected wildlife habitats. This also endangered the existence of the Everglades, characterized by a wetland ecosystem dependent on a continuous flow of overland flow of water.

Large construction projects for new communities, spurred by the mild winter climate especially attractive to retired persons, have negatively impacted the Everglades and other wetland areas. In places, wildlife and wetland reserves and parks have been established to protect some of these ecosystems. What was also established were transplanted facilities favored by the arriving crowds of northern settlers and tourists—gated communities, luxury hotels, golf courses and tennis clubs, shopping malls, and restaurants catering to the tastes of the northerners. Fish and shellfish harvested along the extensive shoreline and offshore are featured in many of the restaurants. A constant struggle exists between development and environmentalism, although development seems to be winning the struggle.

Aquatic Biomes

Although some authors treat biomes as terrestrial ecosystems exclusively, this text includes them as bona fide biomes. Aquatic biomes[21] can be subdivided into *marine* and freshwater. Transitions exist with terrestrial biomes. For example, a freshwater wetland can be a transition between a grassland and a lacustrine (lake) ecosystem. A salt marsh can be a transition between a grassland and an intertidal marine ecosystem. Among aquatic environments, there is a transition produced when a freshwater stream empties into a saltwater bay. Tides from the ocean can move up the stream's mouth and bring salt water to freshwater environments, either by mixing or by forming a separate layer of different salinity.

Marine

The *marine biome* can be subdivided on the basis of location in the oceans (nearness to land and depth). Open sea, shallow sea or ocean surface, deep sea (abyssal), with some or no solar penetration, continental shelf, intertidal zone, deep ocean bottom, continental shelf bottom, coral reefs (**Figure 7-17**), mangrove swamps, and estuaries are some of these ecosystems.

Biota in the marine biome comprise both plants and animals. Their existence and habitat is influenced by seawater properties, such as salinity, temperature, and depth.

Most of the plants in the marine biome are nonvascular; that is, they do not have circulatory systems. Most are plankton, floating near the sea surface. Some have protective shells of silica or calcium carbonate. Most are unicellular, solitary or in colonies (including seaweed). As with terrestrial plants, they form the base of the food web. Some vascular plants exist at the sea margins, like mangroves and sea grass.

FIGURE 7-17 Coral reef

© Alan Jacobs

© Clara Demiral

FIGURES 7-18 and 7-19 Loggerhead sea turtle, Gumbo Limbo Nature Center, Boca Raton, Florida

Sea animals abound, and include all the major phyla. The only noticeable animals absent from marine environments are *amphibians* (Class Amphibia), which are present only in *freshwater* (some brackish) aquatic biomes, and insects. Sea creatures found only in marine environments include reptiles (sea turtles) **(Figures 7-18, 7-19)**, some mollusks (e.g., squid, octopi), some cnidaria (e.g., corals), echinoderms (e.g., starfish, sea urchins), some fish (e.g., sharks, rays, skates), some land mammals that "returned" to the sea (e.g., whales, dolphins, sea otters, manatees), and some birds (e.g., penguins, albatrosses).

The sea provides transportation and shipping routes, recreation, mineral and petroleum exploration along continental margins, and sand and gravel for construction, as well as a source of seafood. Adverse impacts by humans to the marine environment include overfishing, bykill (taking unwanted species among catches), marine waste disposal, and heat discharge into the sea from the circulation of cooling waters from coastal power plants.

Freshwater

The *freshwater biome* can be subdivided on the basis of flowing or standing water. *Flowing water habitats* include streams and rivers. Standing water habitats include ponds and lakes. Transitions, as mentioned previously, can include some influences from their proximity to the sea.

Plants in freshwater, can be vascular (have circulatory conduits) that have stems and leaves that can extend above the water surface. They can also be nonvascular, such as algae and mosses, and live totally submerged. Transitioning between terrestrial and freshwater-aquatic are the plants along drainage ways living in the so-called *riparian zones*. They help to slow bank erosion, control flooding, and provide homes to semiaquatic wildlife animals.

In streams and lakes there are fish, amphibians (e.g., frogs, newts, salamanders), crustaceans (e.g., crawfish), reptiles (e.g., turtles, water snakes, alligators), insects, birds, and mammals (e.g., beavers) (**Figure 7-20**).

Productivity of the freshwater environment is increased by the presence of nutrients and dissolved oxygen. An increase in nutrients from excess fertilizer runoff into this environment, however, can have a deleterious effect. Excess nutrients may cause algae to grow to great depths, cutting the light penetration (preventing photosynthesis) to the lower algal mat.

These algae die, and their decomposer bacteria use up available dissolved oxygen. This condition, called *eutrophication*, produces an oxygen deficiency in the area below the mat. Game fish and other aquatic life cannot survive in these dead zones.

Freshwater environments are sites for fishing, recreation, water supply, hydroelectric power stations, and transportation and shipping routes (on large rivers and lakes). Adverse impacts by humans to the freshwater environment include industrial discharge of pollutants, draining of wetlands, damming of streams, pretreatment sewage discharge, algal blooms, and acid mine drainage.

FIGURE 7-20 Freshwater pond and swan, Singapore Botanical Gardens

Succession

The biomes do not necessarily remain unchanged during the life of the ecosystem.[22] The soil, the land surface itself, the climate, the available moisture, all, can change on a dynamic Earth. When a catastrophic event occurs, the ecosystem adjusts to the changes in a series of "repairs." The repairs can create intermediate environments, themselves unstable, which can be replaced, successively, until a dynamic equilibrium is reached.

For example, a volcano erupts and covers all productive soils with hard lava rock. The soil and vegetation below is burned and buried. Wildlife dies or emigrates to areas not affected by the volcanic eruption. Slowly, the lava rock cools and becomes weathered, primitive soils form, and pioneer vegetation establishes itself and forms local microclimates. These microclimates welcome other plants in a series of stages. Animals find temporary habitats here. Both plants and animals are replaced as the biome recovers.

If recovery leads from a completely barren environment, this *primary succession* can be slow. If the disturbance is only partial, then the ensuing *secondary succession* can proceed more rapidly. What is the end point to the series of changes in this succession? Theoretically, the land will return to the biome from which it had been in equilibrium prior to the disturbance. The community to be reestablished is called a *climax community*. The Earth is dynamic, so there may be differences between the original community and the climax community.

Natural or man-made destruction of ecosystems, fortunately, is not permanent. Through the process of succession ecosystems can reestablish themselves through a series of stages promoting immigration of pioneer species to climax species of plants and animals until the ecosystem is again in dynamic equilibrium with its environment. The Earth figuratively heals itself. This concept, called *Gaia* after the Mother Earth deity, represents hope for damaged ecosystems.[23]

During the 1960s, when the environment was being continuously degraded, it was believed that Lake Erie was dead, without any hope of recovery. Yet, with limitations to continuing pollution, the lake has recovered. Environmentalists still argue that there is a point at which there would be no hope of recovery even if pollution was curtailed. The key element to recovery here is pollution prevention, which should be the rule instead of the exception.

References

1. Renger, G., ed. 2008. *Primary Processes of Photosynthesis: Principles and Apparatus.* Cambridge, UK: RSC Publishing.
2. Steward, F. C. 1960. *Photosynthesis and Chemosynthesis.* New York: Academic Press.
3. Brooker, R. J. 2014. *Biology.* New York, NY: McGraw-Hill.
4. Mullan, D. J. 2010. *Physics of the Sun: A First Course.* Boca Raton, FL: CRC Press.
5. McCann, K. S. 2012. *Food Webs.* Princeton, NJ: Princeton University Press.
6. Withgott, J. and M. LaPosata. 2012. *Essential Environment: The Science Behind the Stories,* 4th ed. Boston: Pearson.
7. Werner, D. and W. E. Newton. 2005. Nitrogen *Fixation in Aagriculture, Forestry, Ecology, and the Environment.* Dordrecht, Netherlands: Springer.
8. Woodward, S. L. 2009. *Introduction to Biomes.* Westport, CT: Greenwood Press.
9. Ward, D. 2009. *The Biology of Deserts.* New York: Oxford University Press.
10. World Book, Inc. 2011. *The Great Depression.* Chicago: Author.
11. Kalman, B. 2007. *A Savanna Habitat.* New York: Crabtree.
12. Warhol, T. 2006. *Chaparral and Scrub.* New York: Marshall Cavendish Benchmark.
13. Fisher, R. F. and D. Binkley. 2000. *Ecology and Management of Forest Soils.* New York: Wiley.
14. Larsen, J. A. 1980. *The Boreal Ecosystem.* New York: Academic Press.
15. Kuennecke, B. 2008. *Temperate Forest Biomes.* Westport, CT: Greenwood Press.
16. Holzman, B. A. 2008. *Tropical Forest Biomes.* Westport, CT: Greenwood Press.
17. Sexton, C. 2009. *Tundra.* Minneapolis: Bellwether Media.
18. Lewis, W. M. 2001. *Wetlands Explained: Wetland Science, Policy, and Politics in America.* New York: Oxford University Press.
19. McCally, D. 1999. *The Everglades: An Environmental History.* Gainesville, FL: University Press of Florida.
20. Alden, P., R. B. Cech, R. Keen, A. Leventer, G. Nelson, and W. B. Zomlefer. 1998. *National Audubon Society Field Guide to Florida.* New York, NY: Alfred A. Knopf.
21. Woodward, S. L. 2003. *Biomes of Earth: Terrestrial, Aquatic, and Human-dominated.* Westport, CT: Greenwood Press.
22. Enger, E. D. and B. F. Smith. 2008. *Environmental Science: A Study of Interrelationships,* 12th ed. Boston: McGraw-Hill.
23. Lovelock, J. 2000. *Gaia: A New Look at Life on Earth.* Oxford, UK.: Oxford University Press.

Social Sciences

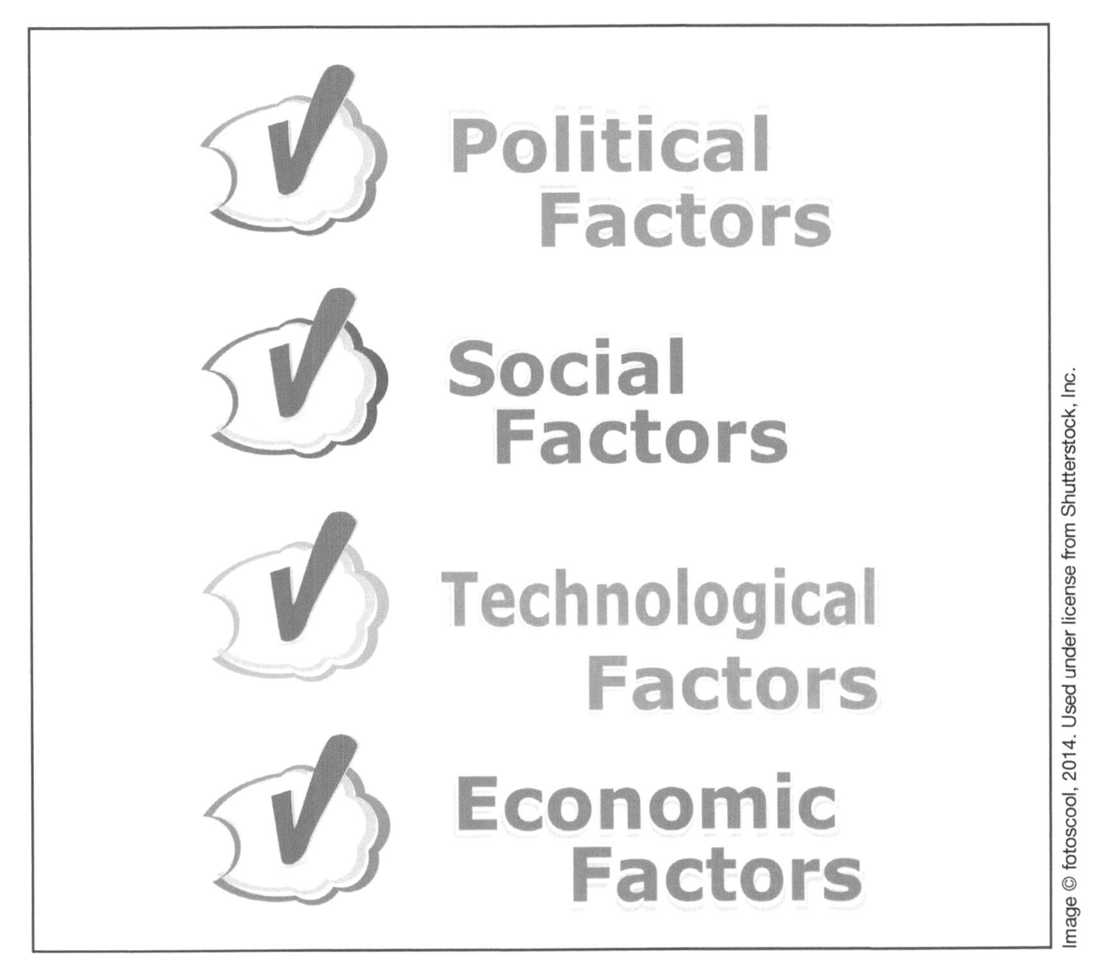

Since 1970, federal, state, and local governments passed significant legislation and developed detailed regulations that address the problems of, and provide solutions to, environmental degradation (see Chapter 27). Non-compliance with these laws and regulations carry severe penalties (fines and in some cases prison terms). Important aspects of these laws and regulations are:

- Liability is strict and several (requires knowledge of the laws and joint responsibility)
- Retroactive (covers actions that took place before the law was passed).
- Due diligence of seller of real estate property to disclose environmental problems before the sale (buyer agrees to liability after the sale).

- Water or soil derived from hazardous waste is considered hazardous
- Prohibits certain wastes or substances from being placed in landfills
- Approves design and permits operation of landfills
- Requires labeling of containers and placarding of transporting vehicles of hazardous materials
- Requires safety training in the handling of potentially hazardous materials
- Requires reporting and permitting of discharges of wastes to the environment
- Regulatory oversight of the assessment and cleanup of contaminated sites
- Procedures for storage, transportation, and disposal of hazardous substances
- Employees have the right to know about hazardous working conditions
- Distinction between municipal solid waste (fewer restrictions) and hazardous waste (more restrictions).
- Requires testing of drinking water by public supplier
- Limits on maximum levels of pollutants in ambient air, water, and soil.

Although regulations were not developed using the scientific method, they are considered "regulatory science." With emphasis on compliance for waste generators, parties responsible for cleanups of abandoned sites, and oversight of providers of safe supplies of water and food, regulations have been developed to be health protective and enforceable, but also technologically and economically feasible and socially acceptable. The result, since 1970, has been a much cleaner environment with greater hopes of sustainable development.

Government can establish a level playing field among waste generators, so that the costs of proper management of wastes do not create a competitive disadvantage in the marketplace. Citizen groups and individual voters can encourage politicians to take the moral and ethical high ground when government considers legislation to protect our environments from abuse.[1]

Economics also plays an important role in environmental sustainability. Costs of environmental protection are offset by better health, market advantages (see Chapter 27) by having a reputation for being a protector rather than a polluter of the environment, a lower cost of raw materials from recycling and reusing materials, and a lower cost of energy to produce goods and services.

If a site is contaminated, regulatory agencies encourage the use of lower-cost cleanup technologies that produce a sufficient reduction in environmental risk for the future intended use of a contaminated site. Programs like *Brownfield redevelopment* provide economic encouragement to companies that voluntarily make efforts to reuse once-contaminated land.

We can have both a safe and clean environment and economic benefits (jobs, profits, and lower costs) if a partnership is maintained between public and private interests.[2]

References

1. Vaughn, J. 1998. *Environmental Politics: Domestic and Global Dimensions*. New York: St. Martin's Press.
2. Thampapillai, D. 2002. *Environmental Economics: Concepts, Methods, and Policies*. South Melbourne: Oxford University Press.

UNIT 3 Sustainable Material Resources

Environmental media include soil (weathered bedrock, sediment, croplands), water (streams, lakes, groundwater, seeps, oceans), waste (discarded materials), or bio-vectors (living organisms that can spread disease). Such media include any part of the environment that might contain or transport nutrients, energy, contaminants, or pathogens (organisms that cause disease). Media can be part of the lithosphere, biosphere, hydrosphere, or atmosphere. In evaluating environmental problems and solutions, media can be the sources, the pathways, or the receptors that might cause or be the recipients of adverse health impacts. Without media to spread pollution and disease, there might not be any environmental problems. However, without media there also may be no life, as the media can contain and transport nutrients and energy to living communities.

Water

Hydrology: How Do We Study Water?

Water, chemically, is H_2O, comprising two atoms of hydrogen and one atom of oxygen in each molecule. The study of water is a subject in various disciplines. Chemistry includes the study of the properties of water and water solutions. Physics includes the study of the physical states of water (solid ice, liquid water, and gaseous water vapor) and water containing isotopes of hydrogen (deuterium and tritium). Water makes up a high percentage of biological tissues and takes part in bodily functions (physiology), habitats of life forms (ecology), and organisms that transmit diseases to humans (toxicology). Water in the atmosphere is studied under the name "*meteorology*" for daily weather phenomena and under the name "*climatology*" for long-term shifts in the climate.

The geology of water, often called "*hydrogeology*," is the study of the movement and quantity of water at and below the Earth's surface. Fresh surface water exists in rivers, lakes, estuaries, wetlands, springs, and seeps. Its study is called "*limnology*." We study the water in oceans, which is saline, under a subheading of "*oceanography*." Water below the surface (groundwater) may be found as soil moisture, in aquifers (permeable water-bearing formations), and in voids (solution cavities and mines). Rivers of ice (valley glaciers) and ice sheets are studied under the subheading "*glaciology*."

Geologists also study the ability of water to chemically weather rocks and soil, running water to erode the landscape, and the transportation in water currents and deposition of waterborne sediment to form new soils, sedimentary rock formations, and landforms (sedimentology petrology, soil science, and geomorphology).

Other subdisciplines of geology also study water. Waters of hydration can exist chemically attached to compounds in certain minerals, and such water is studied under the subdiscipline "*mineralogy*." Water plays an important role in the formation of minerals in igneous and metamorphic rocks (study of petrology).

Engineers trained in the application of science to techniques that improve the environment are environmental engineers. They develop cleanup technologies for wastewater and for contaminated aquifers and surface waters (environmental hydrology).

In environmental science, water is studied based on its sustainable supply and quality, the latter of which is influenced by its ability to transport and recycle nutrients, to carry pollutants and disease causing organisms (pathogens), and to aid in the remediation (cleansing) of contaminated waters. Nevertheless, because environmental science incorporates knowledge from all the sciences (described above), the student of the environment has to be knowledgeable in the above disciplines as it relates to the environment.

Discipline		Study of (relating to water)
Chemistry		Properties of the water and water solutions
Physics		Physical properties of the solid, liquid, and gaseous states of water, and water comprised of isotopes of hydrogen
Biology	Ecology	Water habitats
	Physiology	Water in life forms
	Toxicology	Waterborne pathogens and toxicants
Geology	Hydrogeology	Surface- and groundwater .
	Geomorphology	Landforms modified by running water and glaciers
	Mineralogy	Water of hydration in minerals
	Sedimentology, Petrology	Aqueous deposition of sediment, effect of water on crystalline rock formation
	Soil Science	Weathering, soil moisture
	Glacial Geology	Valley and continental glaciers
	Medical Geology	Geologic materials that can cause or treat disease

Oceanography	Properties of sea water, waves, and currents
Meteorology and Climatology	Water circulation in the atmosphere; glaciation
Environmental Engineering	Wastewater treatment, remediation of contaminated water
Environmental Science	Sustainable water supply, water quality, and pathways of dissolved and suspended contaminants/pathogens plus all the above disciplinary studies.

Occurrence of Water

Most water on Earth, 97.5 percent of it, is salt water in the oceans and in brine solutions deep in the Earth's crust.[1] Of the remaining water, 2.5 percent is freshwater. Of all the freshwater, almost 70 percent is frozen in glaciers, snow banks, or in frozen soils, and the remaining almost 30 percent is groundwater (found in the pore spaces of underground rock and soil). Less than one-half percent of all freshwater is water on the surface and atmospheric water. This seems like a lot of water as we can see mighty rivers, large lakes, and billowy clouds.

But that entire half-percent of the 2.5 percent of all free water on Earth (~1.25% of it) pales in comparison to the combined volume of all other sources of water. The one-half percent of freshwater (here parenthetically called "it") can be divided into water in lakes (2/3 of it), water in soil moisture (1/8 of it), water in the atmosphere (1/10 of it), water in wetlands (less than 1/10 of it), water found in organisms (less than 1/100 of it), and last but not least water found in rivers (about 1/25 of it). Consequently, freshwater that we need is precious.

Water in the Atmosphere

Water in the atmosphere adds to the humidity. The capacity of the air to keep water as a vapor depends on temperature and pressure. Rising air that cools can hold less water vapor, so the likelihood for *precipitation* (e.g., rain, snow) increases. Falling air warms and can hold more moisture, so the likelihood for precipitation decreases.

At higher elevations, water vapor forms clouds of liquid droplets or particles of frozen water. Clouds are the source of precipitation. Their shapes are harbingers of weather events, for example puffy and scattered clouds indicate fair weather, thick layers indicate overcast conditions, and billowy and dark clouds indicate an impending storm.

Precipitation results when the air reaches its capacity to hold moisture in the form of water vapor (100% humidity). At that point water or frozen water droplets can collect around dust particles and fall to the surface as rain, snow, sleet, or hail.

Transpiration is evaporation of water from plants, especially from their leaves. Therefore, plants in arid regions without or with fewer leaves (e.g., cacti and other succulents) can survive in deserts by producing less transpired water.

Added to evaporation from water compartments, the water joins the hydrologic cycle.

Water on the Surface

Precipitation adds to the water at the surface and can run off overland, usually prior to entering channels, bodies of water, and the soil and subsurface aquifers.

The flow of water over land initiates the formation of stream networks by eroding channels. The network keeps the water flowing from tributaries to larger channels. A dynamic equilibrium is maintained by erosion and deposition to create an integrated network of channels from the headwater to the ocean.

Occasionally, the flow of water in streams becomes temporarily blocked by dams, landslides, lava flows, or glaciers and retained in lakes, ponds, or wetlands.

Water reaches the oceans from direct precipitation, inflow from streams, or melting of ice from glaciers. Inflow of these freshwater sources decreases the ocean's salinity. Evaporation of ocean water increases its salinity.

Groundwater

All water below the ground surface is groundwater[2]. It can take the form of soil *moisture*, water combined in minerals or molten rock, and brine in oil pools. When water collects in the pore spaces of soil and rock, it can partially fill the pore spaces in an upper, unsaturated zone and also percolate downward to completely fill the pore spaces in a lower, saturated zone. The upper part of the unsaturated zone is called soil moisture, and is vital to productivity of plants and survival of soil organisms.

There are several misconceptions about water under the ground.

Misconception 1: Water below the ground is found in underground lakes and rivers. Ans. Mostly false.

In areas of caves and caverns, where rock has been dissolved by groundwater, voids are formed to allow water to collect underground in pools and flow through tunnels. Most groundwater, however, is found in tiny pore spaces between grains of rock and soil and in fractures in the rock. Groundwater flow can occur from grain to grain and from fracture to fracture. If the flow of groundwater is sufficient for water supplies, it is called an *aquifer* (water carrier).

Misconception 2: Groundwater is absent under deserts. Ans. False

Precipitation in all landscapes can seep into the ground and can travel downward through rock and soil pore spaces and fractures. It travels first in a zone whose pore spaces and fractures are not all filled with water (the *unsaturated zone*). It eventually reaches the *saturated zone*, where all the pore spaces and fractures are filled with water. In deserts the top of the saturated zone (called the *water table*) is generally deeper than in more humid areas. Where the water table intersects the surface, a *spring* may occur or a river is fed by groundwater. Water tables intersect the surface in deserts at oases.

Misconception 3: To locate groundwater supplies, it is necessary to use a forked stick or L-shaped rods. Ans. False.

There is no scientific reason to expect dowsing tools to sense underground water **(Figure 9-1)**. All subsurface areas (even at greater depths in deserts) have groundwater. Anywhere the dowsing rods seem to dip or cross, or not, in the dowser's hands, groundwater can be found.

Misconception 4: When one drills to install a water well, it is sufficient to drill only to depths where wet soil indicates penetration into the saturated zone. Ans. False.

FIGURE 9-1 Water witch using a dowsing rod (forked stick)

Drilling should continue to greater depths for the following reasons:

1. The time of drilling might be in a wet season, when the water table is shallow. During the dry season, the water table will be lower, causing the bottom of the well to be in the unsaturated zone (unproductive).
2. The saturated zone might be isolated from the regional water table. This can result on an *impermeable*, subsurface shelf of clay or shale. This situation produces a *perched aquifer*, which has a limit supply of water (only temporarily productive).
3. If the bottom of the well is at or below the water table at the time of pumping, the act of pumping, however, will lower the water table locally around the well, leaving the well intake temporarily in the unsaturated zone. This conical shaped depression of the water table is called a *cone of depression* (unproductive). The water table can recover in time after pumping ceases or the rate of pumping lessens.

Aquifers (Porous and Permeable Water Bearers)

An *aquifer (water bearer)* is the term given to rock or soil that has high *porosity* and *permeability*, and constitutes a water supply unit. If a rock or soil unit has a low permeability, on the other hand, it is called an *aquiclude* (excludes water) or an *aquitard* (retards the flow of water).

Rock or soil contains solid particles and pore spaces between the particles. Porosity is the percentage of volume of pore spaces to the total volume of rock or soil. For example, if there is one cubic meter of rock and it has 25 percent porosity, 75 percent of the rock contains solid particles and the remainder just pore spaces. Another measure of porosity, used by engineers, is void ratio. That is the ratio between pore spaces and solid particles. So the same rock described above would have a void ration of 0.25 cubic meters divided by 0.75 cubic meters or one third. If the rock is not porous but has many fractures or solution channels, it can have what is called *secondary porosity*.

The percentage of pores, voids, or fractures in rock or soil is not definitive in determining the ease of flow of water or other fluids through rock or soil. The pores must be connected in order for flow to occur through the rock or soil. Sand and gravel each has high porosity and high permeability, allowing this loose, uncemented material to hold and easily transmit water. Shale, a clayey compacted rock, conversely, has a high percentage of pore spaces (between the layered molecular structure of clay minerals) resulting in a high porosity, but a very low permeability.

Permeability, the ability of the rock or soil to transmit water, is a very important property in water supply or movement of subsurface water as a transporting medium (moving nutrients or contaminants). If the rock or soil is not permeable it is said to be *impermeable*.

Permeability depends on the properties of the rock or soil, rather than also on the properties of the fluid passing through it. Akin to permeability is *hydraulic conductivity*, which also helps to define the properties of the waters (or other fluids) that pass through the rock or soil. Hydraulic conductivity depends on the permeability of the rock and soil and the viscosity of the liquids passing through it. Viscosity varies with the kind of liquid and the temperature of that liquid. Crude oil has a higher viscosity than water; so does honey and molasses. Cold water has a higher viscosity than warm water. Liquids with higher viscosities have lower hydraulic conductivities when passing through the rocks of the same permeability.

The presence of permeable rocks and soils and fractured bedrock determines the availability of groundwater supplies from both unconfined and confined aquifers. Areas of thick permeable soils (sands and gravels) are present in glaciated areas (e.g., Midwest United States) in buried bedrock valleys. These valleys were carved out and filled by glacial melt waters from the Pleistocene Epoch (ice ages). Regionally extensive sandstone formations whose recharge areas are at relatively high elevations are also good sources. Such bedrock aquifers include the *Ogallala Formation* supplying the Midcontinent United States and the Cambrian–Ordovician bedrock aquifer system supplying areas of the upper Midwest United States). Groundwater supplies are important, especially where surface streams and lakes (or man-made reservoirs) would be subject to high evaporation rates, as in arid areas (Southwest United States).

Where an aquifer intersects the ground surface, it can discharge water to seeps, springs, channels, and lakes. Streams accepting groundwater in this manner are called *gaining streams*. Surface streams that recharge water to aquifers are called *losing streams*. Groundwater that is heated by subsurface igneous activity can be heated (or superheated under pressure) and discharged at the surface as a *geyser*.

There are two main types of aquifers, unconfined and confined **(Figure 9-2)**. In an *unconfined aquifer*, the water table is free to move up or down, depending on the quantity of water entering the saturated zone. In rainy seasons, the water table can move up and in dry seasons the water table can move down. In moving up, it can reach the ground surface and form a wetland or on steep slopes it can form a spring or seep.

A confined aquifer is one that has an impermeable layer of soil or rock above it. Consequently, the top surface of the saturated zone cannot rise above the impermeable layer; it becomes confined and pressurized.

The top surface of an unconfined aquifer is the water table. Above the water table is a zone that is not saturated with water; i.e., the pore spaces are not completely filled with water (although they do contain some soil moisture). Water infiltrating from the surface passes through this zone (called the *vadose zone* or the *phreatic zone*) on its way by the force of gravity down to and below the water table. The more water infiltrating down can raise the water table. Below the water table is the aquifer, whose pore spaces are completely filled with water. If water is pumped out or flows out of the aquifer, the water table can be lowered. This aquifer is called unconfined, because there is no layer of impermeable rock or soil (aquiclude or aquitard) above it that restricts the rising or lowering of the top of its saturated zone. Therefore, in rainy seasons the water table rises; the aquifer is recharged. In dry seasons the water table falls. If a boring (or well) extends down into aquifer, water rises in the borehole or well to the level of the water table.

Confined Aquifers

The second type of aquifer is the confined aquifer, which has a confining layer (aquiclude) above the aquifer. Because infiltrating water in the zone above the confined aquifer cannot penetrate down through the aquiclude, it cannot be locally recharged with new infiltrating water.

It is possible to recharge a confined aquifer where the aquiclude is

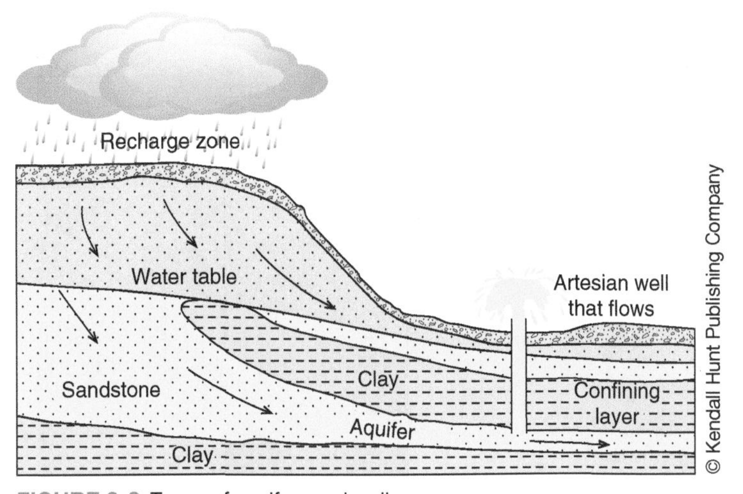

FIGURE 9-2 Types of aquifers and wells

not present, i.e., at locations horizontally distant where the water can flow down and laterally into the aquifer. Also, water cannot rise in the confined aquifer above a nonbreached confining layer. Pressure, therefore, can build in the aquifer when enough water is forced laterally beneath the confining layer, forming an artesian situation. If a boring (or well) extends down through the confining layer and into the confined aquifer (breaching the confining layer), water can rise in the borehole or well to the level higher than the confining layer (artesian well), if there is enough pressure built up. If there is enough pressure to raise the water in the well to the surface, a free-flowing well is possible (without the need to pump water to the surface).

Direction of Groundwater Movement in Aquifers

Movement of water in aquifers is determined by the *hydraulic gradient* (high pressure to low pressure). In unconfined aquifers, the slope of the water table determines this direction. In confined aquifers, the water flows down a piezometric surface, descending from the recharge area. Movement direction is important to know when the aquifer is contaminated because one could then anticipate which wells down gradient might be affected. *Contamination* could be from failing septic systems, underground storage tanks (USTs), abandoned wells, waste dumps, feedlots, agriculture chemical runoff, deicing salts, etc.

Perched Aquifers

Water can collect in a mound-shaped saturated zone over the confining layer. This rather limited supply is called a perched aquifer, since it sits perched over the regional water table. Perched aquifers can fool water-well drillers into thinking that they have reached a regional aquifer. Regional aquifers can continually supply water from flow from adjacent recharge areas. As perched aquifers are not connected to the regional aquifer, they cannot be continually recharged. Consequently, perched groundwater mounds cannot be relied upon for a continual supply of water.

The Hydrologic Cycle

Most of our freshwater comes from groundwater wells and surface reservoirs. Some of this water, however, may not be suitable for human use. Water can be polluted by pathogens, chemical wastes, acid rain, sewage, stagnant (devoid of dissolved oxygen), too deep for economical extraction by wells, or too distant for economic transport to consumers. An environmental service is made available by recycling unsuitable water into suitable water with the help of the hydrologic cycle **(Figure 9-3)**.[3]

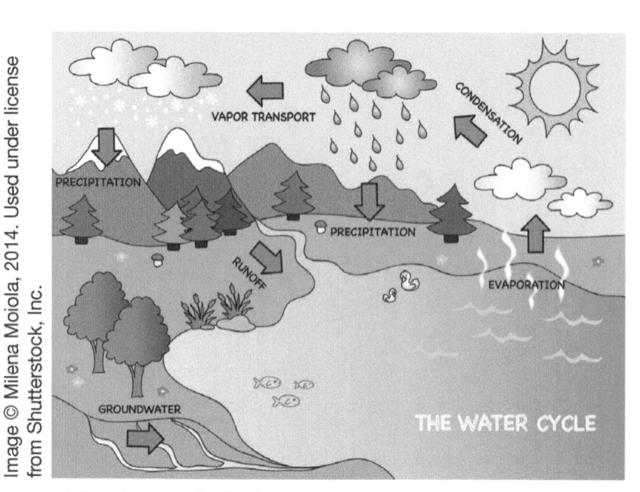

FIGURE 9-3 The hydrologic cycle

The hydrologic (or water) cycle depicts all waters on Earth and is basically a closed system. We do not add water to this system from outside Earth, except for an occasional comet (chunk of ice) that crashes to Earth. We do not lose any water from Earth because of the gravitational attraction of the Earth on all matter. The water, although retained in water compartments (oceans, lakes, rivers, glaciers, etc.), for various periods of time, moves from one compartment to another.

The cycle can be described as follows. Precipitation falls onto the land where it runs off, infiltrates into the ground, or evaporates back into the atmosphere. In the runoff process, it can be trapped in lakes or wetlands, flow in stream channels or overland, and eventually empty into the oceans. In all its travels it can evaporate, infiltrate, freeze and thaw, be taken up by plant roots, and ingested by animals. In the ground, it can move to seeps, springs, rivers, and wetlands.

The Importance of Water

Water is vital to life.[3] We can live without food for several weeks, but we cannot live without water for more than a few days. Water we ingest can come from drinking plain water or water-based beverages, or by eating foods that contain water (most of them do). Our access to pure water is limited, although we need pure water for drinking, food preparation, and washing. Although we do not need pure water for flushing toilets, we use water that meets drinking water standards in toilets because it would be too costly to bring so-called gray water into our houses and buildings in separate conduits. Water that may or may not be up to drinking-water standards is used for manufacturing chemicals and other products, for cooling electrical power generating stations, and for mining and ore processing. Let us call water that is not bound in rocks or manufactured products "free water."

Water and Civilization

Water has been properly managed and mismanaged on land by humans throughout history.[4] Deposition by floodwaters can enrich soils with mineral matter. The Fertile Crescent of the Middle East along the Nile and the Tigris/Euphrates profited from this enrichment during the historical development of agriculture, a milestone in the development of human civilization. Present day river basins, when not constrained by artificial levees, dams, and channelization, can benefit from floods on floodplains. Protection of homes from flooding on floodplains is a costly process; homes should not be built on floodplains.

Water as a Habitat

Aquatic life lives and obtains its resources in water.[5] Some plants float on the water (phytoplankton), are totally submerged (seaweeds), or have only their roots in salt or brackish water (mangroves). Aquatic animals float (*zooplankton*), swim (fish, crustaceans, mollusks, otters, seabirds), or are anchored to the bottom (anemones, crinoids, sponges). They use gills to extract oxygen that is dissolved in water (fish, crustaceans, mollusks), or use lungs and go to the surface to breathe in oxygen from the atmosphere (reptiles, seabirds, whales, dolphins, penguins).

Do we and other air-breathing organisms breathe air as a gas directly through the lungs and into our bloodstreams? Ironically, the answer is: not exactly. In our lungs, the oxygen is dissolved on the moist inner coating of our lungs, and it passes from water through our lungs into the bloodstream.

Phyla of all living aquatic creatures live and have their origins in water. Some sea creatures ventured onto the land, but retained an aqueous internal environment in their bodies. Amphibians must keep their bodies moist or submerged in freshwater. Aquatic life has adapted to the freshwater environment, and some aquatic creatures (e.g., salmon) divide their life cycle between fresh and salt water.

Water, Including Ice, as a Geomorphic Agent

Water helps to weather rock, to remove sediment, and to redeposit the sediment elsewhere. In the process, water erosion carves landscapes and seacoasts to produce landforms such as buttes, mesas, alluvial fans, sea cliffs, drainage basins, natural levees, meander cutoffs, natural arches, canyons, etc. The study of the formation of these features is called *geomorphology*, a branch of geology and geography.[6]

As a geomorphic agent, water is in dynamic equilibrium with the landscape it is sculpturing. The velocity of the water, its sediment load, and the potential energy it has due to gravity determine whether it erodes or deposits and what size particles it can carry or drop.

Water as a Pathway for Contaminants and Pathogens

Water can serve as the pathway to deliver contaminants from their sources to receptors and cause adverse health effects.[7] Water can dissolve toxic metals from the soil (e.g., lead and arsenic) and add them to water supplies. *Leachate* from waste can also mix with pure water to pollute water supplies.

Although water is a habitat for aquatic creatures, it is also an incubating and transporting medium for *waterborne pathogens*, such as bacteria, viruses, and protozoans. Animals that ingest pathogens in the water can contract gastrointestinal diseases. Larvae of mosquitoes, whose female adults are common disease vectors, breed in water.

Water as a Purifying Agent

When movement through the hydrologic cycle involves a phase change to water vapor from liquid water (*evaporation*), from solid ice (*sublimation*), or from water in plants (transpiration), the impurities usually are not incorporated into the water vapor. This is like a distillation process, which leaves impurities behind and concentrates the fresh, pure water.[8]

During this movement, pure water can dilute contaminated water and disperse contaminants. Oxygen that is dissolved in the water can be robbed by respiring organisms, or replenished by aeration in contact with the atmosphere, especially during turbulent flow. Groundwater can be filtered by porous media through which it flows. Contaminants can adhere to soil or sediment particles. Chemical reactions can take place, transforming contaminants into less (or more) hazardous solutes.

Obtaining Water for Use

We can extract fresh, clean water from certain surface and groundwater sources.[9] It can be artificially retained in cisterns, rain barrels, water towers, etc., for later use. Surface water flows downhill, so if we need water delivered to higher elevations it must be pumped. Groundwater can be obtained by pumping. In certain situations, an artesian well tapping a confined aquifer can provide water with little or no pumping.

Wells (Drilled into Aquifers)

A well is a drilled or dug device used to extract fluids from underground sources, to inject waste liquids, or to measure aquifer parameters (e.g., aquifer permeability, water quality, depth to the saturated zone). When wells are just to measure the depth to the saturated zone, they are called *piezometers* and are usually less than two inches in diameter.

Installation starts with the drilling of a borehole down to the proper depth or horizon and then installing pipe or other materials and equipment to satisfy the intended use. If the establishment of a water supply is the intent, usually, casing for the borehole, sealant in the annulus between casing and borehole, screens and gravel to filter the water, and submersible pumps are installed. *Boreholes* can be advanced by: augers, cable tools (ramming a cutting tool), and rotary drill-bit systems depending on depth, resistance of strata, and need to take samples (cuttings, split-spoon soils, or undisturbed cores of rock).

Pumps are installed at the bottom of well bores. That water enters the water well bore and is forced to the surface by submersible pumps that "lift" the water by pushing it from below. In unconfined aquifers, pumping lowers the water table. Seasonal fluctuations in the water table (wet and dry seasons) are common. In the saturated zone, groundwater flows down gradient (down the slope) of the water table. Air then temporarily occupies the pore spaces or cracks in the aquifer creating a conical shaped unsaturated zone, called a *cone of depression*. The cone depresses the water table, and keeps depressing it, as pumping continues. If the base of the cone or the regional water table falls below the bottom of the bore hole, then the well dries up and is useless. In time, the cone of depression is naturally refilled with water flowing from the surrounding aquifer (recharge). The rate of recharge depends on the permeability of the aquifer, the availability of water in the aquifer, and the slope of the water table.

Extensive pumping that significantly alters the slope of the water table can actually reverse the flow of water in the aquifer. With the water flow, contaminants in the water can be carried along in the same direction. Therefore, extensive pumping can draw contaminated water to drinking water wells from sources of contamination up gradient. Conversely, when contaminated water flows in the direction of existing water wells, pumping up gradient from the source of contamination can help to draw the contaminants away from the drinking water wells.

Water in the aquifer should be tested for purity. Furthermore, the annulus of the well (borehole sides) above the aquifer and the well head should be sealed, so that only the aquifer is tapped, rather than waters seeping into the borehole above the aquifer. Sealing prevents contaminants in the surface soil or in shallower, contaminated aquifers from leaking into the well.

Water Use

We use water to drink, to cook with, to wash, to flush toilets, to irrigate our crops and gardens, to put out fires, to cool and heat by water circulation, to add to manufacturing processes, to fill swimming pools, aquariums, farm-raised fish ponds, as a solvent to dissolve and dilute solutes (e.g., salt, sugar, nutrients in our bodies), to keep dust down on unpaved roads, make ice cubes, generate steam to turn turbines in electrical generating power plants, to help swallow medicinal pills, to brush our teeth and gargle, and to steam-clean objects. We need water for these uses, but we also waste water when we let it run or drip from the tap without efficient use. We use water cleaned to drinking water standards to flush toilets. We sometimes allow sewage to invade our well-water supplies by improper wastewater management. We need an adequate supply and the supply must include water of sufficient quality for its intended use.

Water Quality

If we ingest water (beverages and foods) the water must be maintained at drinking water standards. It must be tested periodically. It must be free of sewage contamination, which may contain pathogens. Proper standards for public water supply systems according to the Clean Water Act include concentrations of chemicals in the water at or below *maximum contaminant levels (MCLs)*. MCLs vary among the possible contaminants of concern. Contaminant concentrations above MCLs of lead, arsenic, mercury, and organic chemicals, when detected require the water supplier to notify consumers. Sewage indicators like the presence of *coliform bacteria* (bacteria from animal/human digestive track) require the supplier to notify and prevent consumption of the contaminated water. When protozoans are found in the supply, it may be necessary for water users to boil the water, especially if chlorination is not effective.

Standards for nonconsumptive use are less strict, yet monitoring of water in swimming pools or public baths is important to maintain public health.

For consumption and personal use, public water suppliers in developed countries purify the water to high standards. Various processes are used depending on local water quality issues.[10] These processes include adding chemicals to precipitate and agglomerate solids (e.g., alum), settling out solids, passing the water through sand beds or membranes, adding disinfectants (e.g., chlorine), adjusting the pH, irradiating using ultraviolet light, bubbling ozone through the water, passing the water through activated-carbon beds, softening the water (removing dissolved calcium compounds), improving the taste or odor, and removing dissolved iron and manganese compounds (that might stain laundry and sinks/tubs).

Nevertheless, the general public may choose to buy bottled water for drinking, which may circumvent some of these purification processes, including the addition of chlorine as a disinfectant. Others reuse the plastic bottles to carry around unrefrigerated water during the day. They might refill their bottles after passing tap water through an activated carbon filter (e.g., *Britta*®). This removes the chlorine and some of the chlorinated hydrocarbons found in tap water, which improves the taste. However, left unrefrigerated and unchlorinated, bacteria and other pathogens may grow in the bottle.

Public water supplies are monitored by law (*EPA regulations*) for pathogens and many chemicals. Alerts to the general public are issued when waterborne disease outbreaks occur. Then, the public is directed to boil water or seek alternate supplies until the danger has passed. The main problem with water supplies is their potential contamination with sewage. Therefore, water (for public supplies) is tested daily for the bacteria that come from the intestines of humans and other animals. These organisms are called *coliform bacteria*. They are not necessarily dangerous to humans, but their presence is an indication of fecal contamination that may also contain pathogens shed from infected humans and animals. Therefore, any coliform bacteria in drinking water supplies are unacceptable.

Regulations that require the monitoring of public *water* supplies stem from the federal *Safe Drinking Water Act (SDWA)* adopted in 1974 with amendments in 1986 and 1996.[11] Most U.S. states and territories set and enforce their own drinking water standards (maximum legal limits and testing schedules), which must be as strong, or stronger, than the federal standards.

It should be understood that the MCLs reflect both the protection of public health and the levels that public water systems can achieve using the *best available technology (BAT)*. In other words, it may be technologically or economically unfeasible to detect contaminants or remove contaminants at very low concentrations, even though the contaminants pose some risk for adverse health effects. As an example, the MCL for *vinyl chloride* is 2 micrograms per liter (µg/L) of water. An "acceptable" excess cancer lifetime risk for a healthy adult is 1 in a million. At this risk level, the concentrations should not be greater than 0.024 to 0.048 µg/L. Unfortunately, the cost of analyzing down to 0.024 to 0.048 µg/L is very high, and is cumulative if frequent testing is required.

Water testing schedules for some chemicals or metals, furthermore, may not allow detection of new contaminants until the next round of sampling. This may be quarterly, yearly, or at longer or shorter intervals depending on whether MCLs have or have not been achieved. As an example, the city of Youngstown, Ohio, in its Drinking Water Consumer Confidence Report for 2010 found no detection of lead or copper above the legal limits, but the sample year listed was for 2008 (two years prior). If the concentration changed since 2008, the results would not be known until the next round of sampling for those metals. As the water company cannot control potential contamination of these metals from the consumers' piping systems, they caution consumers who might have lead or copper plumbing components to test their own water from their taps.

Nevertheless, public water companies test and retest for many contaminants looking for microorganisms, disinfectants, disinfection byproducts, inorganic chemicals, organic chemicals, and radionuclides. Water-company reports to the state or federal EPAs may trigger alerts, prompting retesting more frequently and at lower detection limits.

Cleaning Up Used Water (Waste Water)

Normal usage of water involves some degree of waste incorporation, requiring subsequent cleansing prior to reuse.[12] The cleansing operation is part of water management. Uncleansed water or hazardous materials that contaminate other supplies is called *water pollution*. This pollution can be from a *point-source* (e.g., wastewater pipe (**Figure 9-4**), leaking storage tank, or septic system drain field) or a *non-point source* (e.g., pavement runoff into storm sewers, farm-field runoff, acid precipitation).

FIGURE 9-4 Wastewater discharge

Water can be robbed of dissolved oxygen (DO). A condition called *biochemical oxygen demand (BOD)* results when the DO is depleted by decomposition of dissolved organic matter in the water by aerobic bacteria. A value of 1 milligram per liter of water (1 mg/L) or 1 part per million (ppm) is an ideal value for a BOD. This means that there is only 1 ppm of organic matter measured at 20°C over a 5-day testing period. *Raw sewage* entering a wastewater treatment facility can have a BOD of several hundred ppm and the outflow might be reduced to about 20 ppm. Raw sewage entering a stream with a DO capable of supporting a sporting fish population (about 1 ppm BOD) can increase the BOD causing an "oxygen sag" and resulting in a fish kill in the vicinity of the sewage discharge.

Oxygen dissolved in water is necessary for aquatic life respiration. Fish and other animals use gills to extract the oxygen. *Replenishment of oxygen* can be derived from photosynthesis by plants (mostly algae) and somewhat from the mixing from atmospheric oxygen at the surface of water bodies. The balance between oxygen usage and replenishment can be disturbed by human intervention. Intervention can produce an oxygen sag by the addition of organic waste effluent or eutrophication by the addition of fertilizer runoff.[13]

Sewage (improperly treated or raw), which contains a BOD, when discharged into a stream, can form a food source for bacteria. These bacteria use oxygen to decompose the sewage forming an oxygen sag (low dissolved oxygen concentration) in the river downstream from the point of discharge. As a result, much of the aquatic life, especially game fish, cannot survive here and must travel to water with a lower BOD. The water down or upstream from the oxygen sag has recovered from the high BOD.

A second source of decomposing organic matter results when algal blooms proliferate and form thick mats of algae.

Water receiving an abundance of nutrients from runoff or an overabundance of nutrients from fertilizers on agricultural lands is apt to develop thick mats of algae. As algae must have sunlight to live. When very thick mats of algae shade the lower layers of the mats, algae on the bottom die. At that point, bacteria decompose the dead algae, using up available oxygen beneath the mat. Dissolved oxygen levels decrease and many species of fish may die.[14]

Although public water supplies in developed countries are now relatively safe, this was not the case until the latter part of the twentieth century, and there are still occasional outbreaks of waterborne diseases. During alerts, residents are required to boil tap water (if the contaminant can be destroyed by boiling) or choose alternate sources. In developing countries, waterborne diseases from drinking water are still a major problem. Drinking water purification and sewage treatment are lacking in many areas of Africa and Asia, where raw sewage or defecation of feces may be discharged directly into areas of bathing, washing, and areas of drinking water withdrawal (without purification). The increase in sanitation of our water and treatment of wastewater has played a major role in adding 25 years to our life span since the turn of the twentieth century.

Cooling water discharge from a thermoelectric power plant can return at much higher temperatures. The heated water can cause aquatic wildlife to suffer and waterborne pathogens to grow.

Leaking hazardous products or wastes[15] can enter bodies of water. The hazardous materials can impact aquatic habitats and wildlife. They can also impair water quality at water-supply intakes at lakes and streams and at desalinization plants on seacoasts. Accidental spills from oil tankers and wastes from marine vessels can pollute seawater.

On land, leaks from underground and surface storage tanks (chemicals and processed fuels), septic systems (household wastewater), and surface impoundments (chemicals) can also seep into aquifers and surface waters. Accidents involving vehicles transporting hazardous wastes and products can also cause spills that could pollute water supplies.

To monitor the discharge of pollutants from possible sources, monitoring wells are installed around the perimeters or down gradient from such sources. Monitoring wells are not intended as a source of water supply, but a means of sampling the groundwater and determining the direction of flow of potential contaminants.

Contaminants can be dissolved in the water up to their saturation point. Such contaminants (solutes) are equally distributed throughout the water (solvent) in the aquifer. Above the saturation point, the contaminants separate out of groundwater as *nonaqueous-phase liquids* (NAPLs).[16] High-density NAPLs (DNAPLs) are heavier than water and settle to the bottom of the aquifer (sinkers). Low-density NAPLs (LNAPLs) are lighter than water and rise toward the water table (floaters). Gasoline (perhaps from a nearby leaking underground tank at a filling station) polluting a water-supply well would not dissolve appreciably in the water, but instead would float on top of the water. *Perchloroethylene* (PERC), common dry-cleaning fluid, is a DNAPL and would tend to sink to lower depths of the aquifer. Groundwater supplies containing LNAPLs can be cleansed by skimming the product from the top of the aquifer through wells. DNAPLs are much more difficult to remove; they sink to various depths in the aquifer. Some DNAPLs, those with very high *octanol–water partition coefficients* (e.g., PERC and other chlorinated hydrocarbons) cling to soil particles making them even harder to remove from the water in aquifers.

References

1. Linde, B. M. 2005. *Water on Earth*. Pelham, NY: Benchmark.
2. Nonner, J. C. 2010. *Introduction to Hydrogeology*. Boca Raton, FL: CRC Press.
3. Friis, R. H. 2011. *Essentials of Environmental Health*. Sudbury, MA: Jones and Bartlett.
4. Eder, D. L., ed. 2012. *Aquatic Animals: Biology, Habitats, and Threats*. New York: Nova Science.
5. Sawvel, P. J. 2008. *Water Resource Management*. Detroit, MI: Gale, Cengage Learning.
6. Charlton, R. 2008. *Fundamentals of Fluvial Geomorphology*. New York: Routledge.
7. Watts, R. J. 1998. *Hazardous Wastes: Sources, Pathways, Receptors*. New York: Wiley.
8. Fetter, C. W. 2001. *Applied Hydrogeology*. Upper Saddle River, NJ: Prentice Hall.
9. Kenny, J. F. et al. 2005. *Fact sheet: Estimated Use of Water in the U.S. in 2005*, USGS Circular 1344.
10. Yuncong, L. and K. Migliaccio, eds. 2011. *Water Quality Concepts, Sampling, and Analyses*. Boca Raton, FL: CRC Press.
11. U. S. Environmental Protection Agency. 1996. *Drinking Water Regulations and Health Advisories*. Washington, DC: Author.
12. U.S. Environmental Protection Agency. 2004. *Primer for Municipal Wastewater Treatment Systems*. Washington, DC: Author.

13. Spellman, F. R. 2011. *Spellman's Standard Handbook for Wastewater Operators,* 3 vols. Boca Raton, FL: CRC Press.
14. Webber, C. D. 2010. *Eutrophication: Ecological Effects, Sources, Prevention and Reversal.* New York: Nova Science.
15. Lehr, J. et al. 2002. "Oil Spills and Leaks." Chap. 9. In *Handbook of Complex Environmental Remediation Problems.* New York: McGraw-Hill.
16. Yong, R. N. et al. 1992. "Contaminant Transport Modelling: Transport of Non-Aqueous Phase Liquids." Chap. 9. In *Principles of Contaminant Transport in Soils.* New York: Elsevier.

10 Air

Sustainable Clean Air Supply

Air is an environmental resource. We, and most creatures, rely on it for the supply of oxygen. Creatures with lungs can extract oxygen from the atmosphere. Insects have a system of tubes used to introduce oxygen from the atmosphere into their bodies. Creatures with gills can extract oxygen that is dissolved in water.

Our consumption of nitrogen compounds in our food for protein synthesis comes from atmospheric nitrogen fixed by bacteria that live in the soil at the roots of leguminous plants. Air allows flying birds and insects (and humans in airplanes) to travel aloft above the Earth's surface. Pressurized air provides a smooth ride in tires and buoyancy in aquatic vehicles. We use circulating air to carry heat around our food (convection ovens) and in our living quarters (forced air furnace systems). Circulating air bathes us in a blanket that provides a uniform temperature on all sides of our body. Our most important exposure to air, however, is air in our breathing zone. A sustainable supply of clean air to breathe is vital to our existence.

Air that was laden with particulates and gases from smoke stacks was common in most of the twentieth century. In the early part of that century, street lights in industrial cities were lit in the middle of the day because air pollutants clouded the sky, blocking sunlight from penetrating to street level. We could see and smell the pollution. Office workers in cities brought a change of clothes to work as their clothing by noon was coated with soot. Even spotty locations in rural areas had air that was polluted with power plant or mill stack emissions.

Some equated air pollution with a robust, healthy economy. Stinky air meant jobs and a thriving industry. If signs of hard work included sweat, then signs of prosperity included pollution. Yet, the connection between air pollution and respiratory and other diseases was not considered, so health-care costs and death were not factored into the perceived prosperity concept. Most adults (including physicians) smoked cigarettes, cigars, and pipes, so what was the harm of a little atmospheric smoke added to the mix?

Since passage of Clean Air Act legislation and strict compliance with many of its regulations, we have improved air quality. Many people have given up the addictive habit of smoking, or have been forced to confine smoking where secondhand smoke does not pollute the air of others. Automobiles in industrial areas must be checked for tailpipe emissions that may be spewing high levels of carbon monoxide, volatile organic compounds, and sulfur and nitrogen oxides (all unhealthful air pollutants), and their engines must be tuned to reduce those emissions to acceptable levels. Stack particulates and gasses are being controlled by adjusting combustion processes and precipitating pollutants prior to their release to the atmosphere.

With our increased population and the resulting need to increase industrial output, political pressures have appeared that favor reduction in enforcement of environmental regulations. Such relaxation of the intent of the Clean Air Act, a sustainable, clean air supply, will compromise our health and our real prosperity.

Our Atmosphere

An envelope of gases surrounds the surface of the Earth and extends to a few hundred miles above the surface. Air concentrations are greatest near the ground surface, and progressively decrease at higher elevations, until a complete vacuum is reached at about 400 km (~250 miles) high. These gases comprise 78 percent nitrogen, 21 percent oxygen, 0.9 percent argon, 0.04 percent carbon dioxide, and traces of other gases **(Figure 10-1)**.[1]

Importance of Oxygen in the Atmosphere

All living terrestrial (nonaquatic) creatures (except for certain bacteria) need the oxygen in air to live. It is the basic ingredient for respiration, the process we use to extract energy from food and synthesize with other constituents (nitrogen compounds, water, and minerals) the building blocks of our bodies. Our range of tolerance for oxygen concentration is about what is contained in air near ground level (~20 percent). Less than 17 percent oxygen in air would be lethal. Greater than 25 percent oxygen in air would be hazardous by increasing the flammability of combustibles.[2]

Oxygen content in the air at high elevations might need augmentation. The first sign of altitude disease is a giddiness or faintness. Mountain climbers often carry canisters of oxygen. Pilots, their passengers, and astronauts must have supplied oxygen when flying at high altitudes, where the atmosphere is thin or absent.[2]

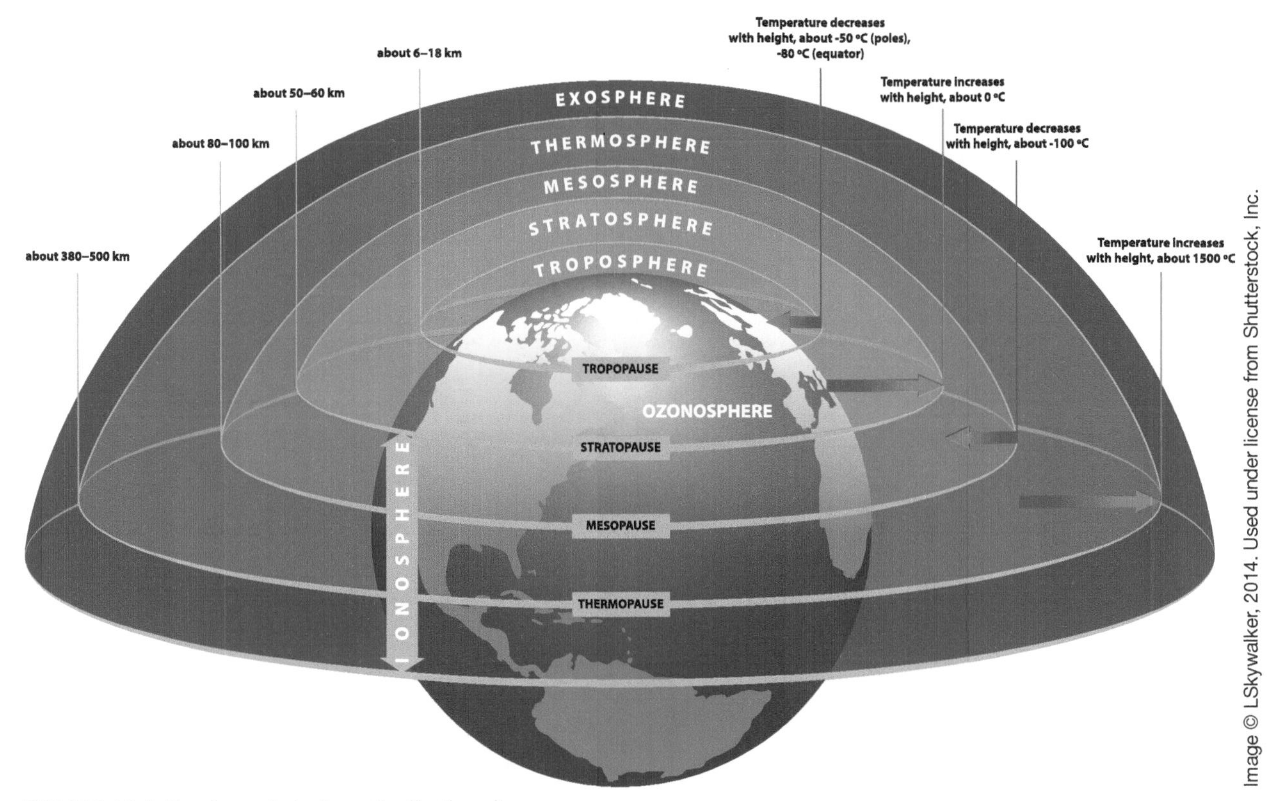

FIGURE 10-1 Envelope of air above the Earth surface.

Oxygen content below ground level (in confined spaces) or at ground level (in low-lying areas) may also be inadequate for life. Oxygen can be displaced by other heavier gases (e.g., hydrogen sulfide) in confined spaces. Manholes used to service underground utility lines and conduits are especially dangerous with respect to the potential for collecting oxygen-displacing gases. In low-lying areas (topographic basins and swales) in cities, smog can settle near the ground surface and displace oxygen rich air. The smog chemicals, in addition to lower oxygen concentrations, are health hazards. Smog may not be dispersed if there is no wind and if there is a thermal inversion (warm air above trapping smog-laden air below).

Layers of the Atmosphere

Because the molecules of all gases in the atmosphere decrease in concentration with height (lower gravitational attraction), the air pressure decreases with altitude. This is a steady decrease from ground surface to the vacuum of space. Temperature fluctuations, however, increase and decrease with height above the surface. These temperature changes define the layers (**Figure 10-2**) of atmosphere, as follows.[3]

Troposphere

The lowest layer, the *troposphere*, extends from ground surface to about 18 km (~11 miles). We live and breathe in this layer. Most weather phenomena occur in this layer. Temperature decreases with height.

The troposphere is the most important layer with respect to the environmental medium called air. Air harbors gases and particulates. In motion, it transports oxygen, critical to life. It also transports polluting gases and particulates. In air, *primary pollutants* (discharged directly into the air) can with available water vapor and sunlight be chemically changed to *secondary pollutants* to form smog. Because the Sun warms the troposphere

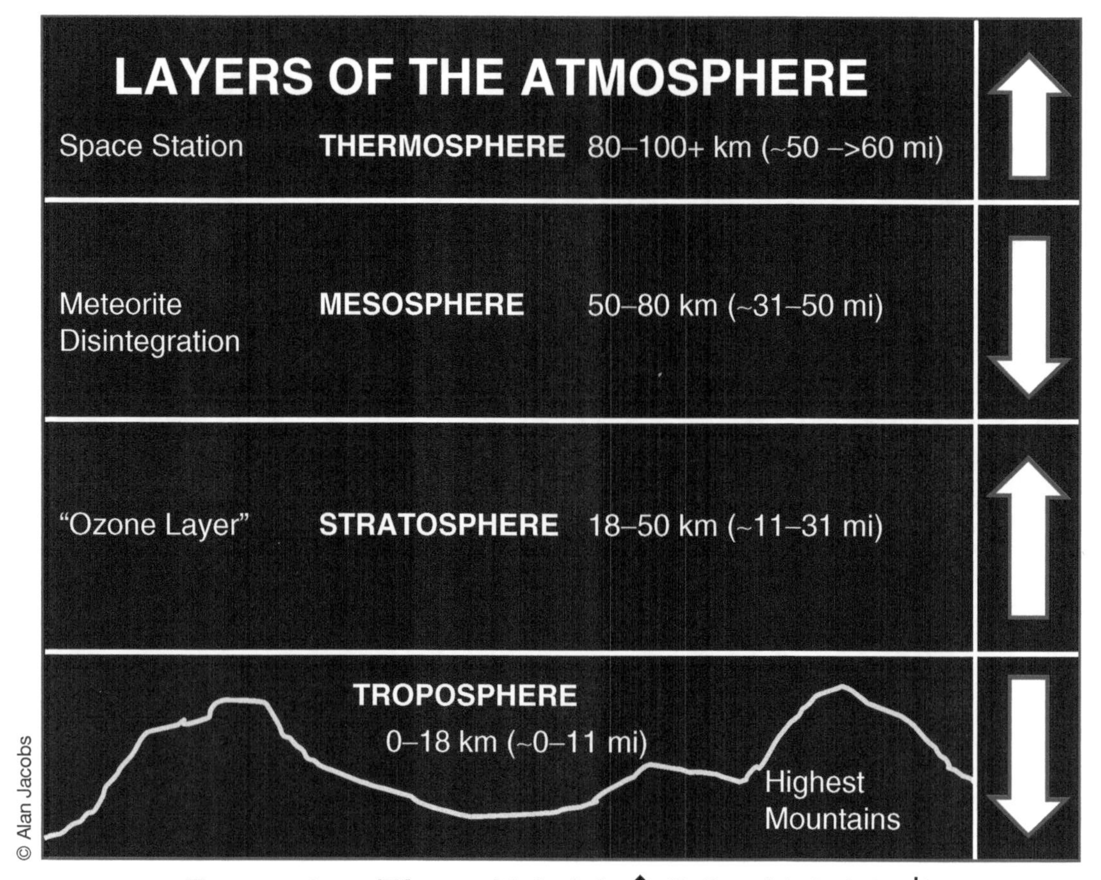

FIGURE 10-2 Layers of the atmosphere

gases unequally, air currents develop, which move air and its contents horizontally and vertically.

When air moves upward, it can cool and hold less moisture and cause precipitation. Precipitation can cleanse the air of particulates, but can also dissolve gases that can produce acid rain. Where air moves horizontally, it can carry its contents across the globe and mix with air of different temperatures and gas concentrations.

Stratosphere

The second layer, the *stratosphere*, extends from the top of the troposphere to about 50 km (~31 miles). Here lies the famous "*ozone layer*," where air contains a greater concentration of ozone gas, which helps to filter out harmful UV radiation. Temperature increases with height.

The stratosphere contains ozone, a triatomic form of normal oxygen (which is diatomic). Ozone can absorb UV radiation, protecting our skin, genetic cells, and eyes. However, UV radiation is not completely absorbed by stratospheric ozone. Because of the health dangers of UV radiation, one should avoid basking in the Sun or the use of artificial UV radiation (tanning beds).

Mesosphere

The third layer, the *mesosphere*, extends from the top of the stratosphere to about 80 km (~50 miles). Temperature decreases with height. The top of the mesosphere is the coldest part of the Earth, reaching about 140° below zero (Celsius). Environmentally, this is the atmospheric layer where most meteorites disintegrate. Collisions with air particles prevent these projectiles from space from hitting the Earth's surface. Thus, aircraft fly below this layer.

Thermosphere

The fourth layer, the *thermosphere*, extends from the top of the mesosphere to more than 100 km (>60 miles). This layer is the site of the Aurora borealis (northern lights) and the Aurora australis (southern lights). These phenomena are caused by solar radiation ionizing particles in the atmosphere. Temperatures increase with height. The International Space Station orbits in this layer.

Although the temperature is high in the thermosphere, the total quantity of heat per unit volume is low. Temperature is the average kinetic energy of the molecules (here they are sparse). Averages can be misleading. For example, in a game of darts, two darts could land across from each other on the outer circle of the target. The vector average is a bull's eye! The air is so thin in the thermosphere that contact with the few air particles may not be sufficient to "feel warm." Also, bodies in this layer might burn up facing the Sun and freeze facing away from the Sun. Solar radiation can transmit heat in a vacuum, but the thin air cannot bathe the body in an envelope of warmed air.

Beyond the thermosphere is outer space, the *exosphere*.

Air Circulation

The Earth orbits the Sun in an elliptical path. Its polar axis is not perpendicular to the plane of this ellipse, but is at an angle of about 66½ degrees from that plane (**Figure 10-3**). We define the axial tilt as its complementary angle, or 23.5° (90–66.5). In the figure, we might call the axial tilt, the angle from the vertical—but, in this situation, what is vertical? During the Earth's orbit, the poles, in their turn, are pointed towards the Sun in their summertime, and away from the Sun in the wintertime.

Because of the axial tilt, solar energy reaching different parts of the Earth's surface and in different seasons is either concentrated when the rays impinge onto the surface more directly (near perpendicular), or more spread out when the rays enter at angles much less than 90° (skimming the surface). This causes a differential heating of the Earth's atmosphere.

Reacting to the differential heating, warm air masses rise and cold air masses descend. Theoretically, the rise of air is most noticeable at the Equator and the descent of air is most noticeable at the poles. What happens, however, is the development of three circulatory cells in each hemisphere (north and south). From poles to Equator, we have the Polar, the Ferrel, and the Hadley cells (**Figure 10-4**). The revolving of the Earth on its axis also generates horizontal movement of the air and imparts a curve of movement, called the *Coriolis effect*.

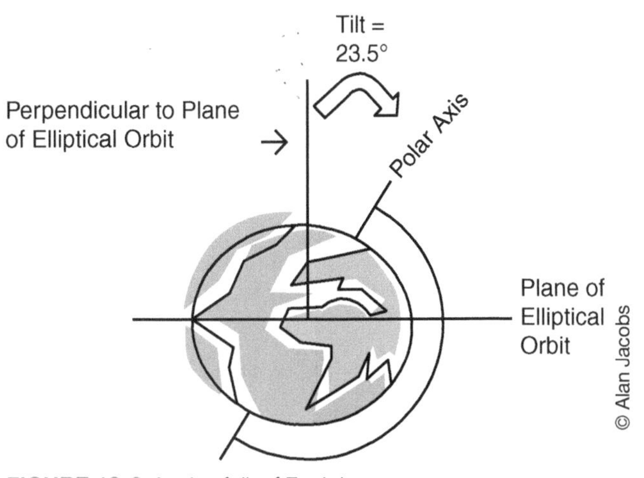

FIGURE 10-3 Angle of tilt of Earth in space

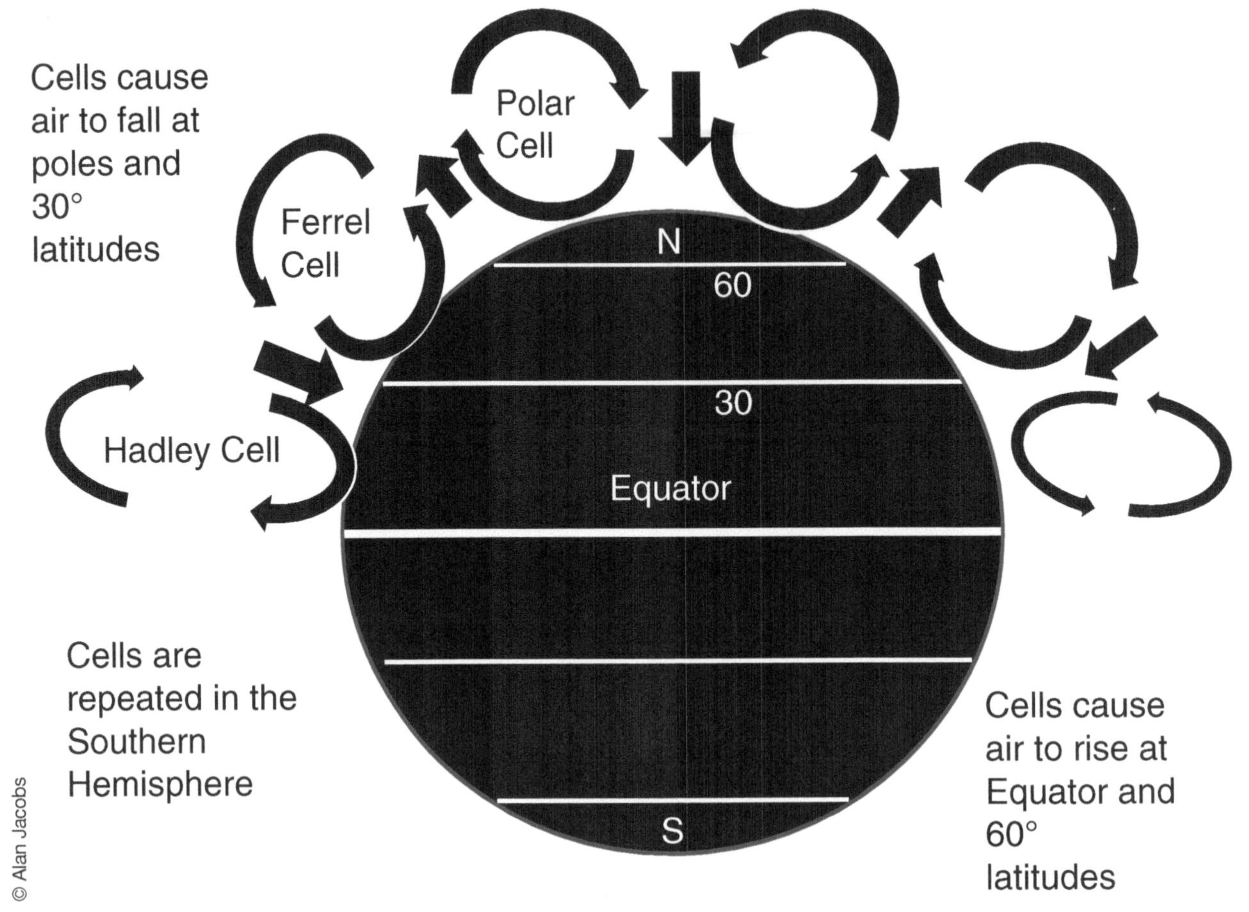

FIGURE 10-4 Atmospheric circulation cells

When two cells converge and channel air currents towards the surface (Hadley and Ferrel, Polar) low-precipitation (desert) conditions are produced at 30° N and 30° S latitudes and at the poles. When two cells converge and channel air currents upwards (north Hadley and south Hadley, Ferrel and Polar) high-precipitation conditions are produced at 60° N and 60° S latitudes and at the Equator.

All these factors produce a predictive model of atmospheric circulation (with some exceptions). The model helps us to understand weather phenomena and their effect on movement of water vapor in the atmosphere.

The exceptions to the theoretical model of atmospheric circulation are a result of differences of heat capacity of oceans and lands (explanation of monsoons), topographic influences of air flow on land (rain shadows), and influences of ocean currents on the atmosphere (El Niño effects).

References

1. Wells, N. 1998. *The Atmosphere and Ocean: A Physical Introduction,* 2nd ed. Hoboken, NJ: Wiley.
2. Occupational Safety and Health Administration. 2012. *Respiratory Protection.* U.S. Department of Labor, Occupational Safety and Health.
3. Ahrens, C. D. 2012. *Essentials of Meteorology: An Invitation to the Atmosphere.* Belmont, CA: Brooks.

11 Soil

Image © TFoxFoto, 2014. Used under license from Shutterstock, Inc.

Sustainable Soil

Managed correctly, soil can be used in agriculture and landscaping, construction, flood control, water supply, and remediation.[1] With the proper nutrients, water supply, and aeration, soil is a growing medium for crops, ornamental horticulture, orchards, and grasslands for grazing animals.

Soil is part of the foundation of construction projects, a raw material used in construction (bricks, concrete, ceramic tiles, caissons, shorelines, and stream bank protection materials). Soils retard overland runoff and increase infiltration of precipitation into groundwater aquifers. Soil aquifers supply water for consumption and irrigation.

Soil is also the habitat for burrowing mammals, insects, and worms. Bacteria in soil can digest organic contaminants. Nitrogen-fixing bacteria can take nitrogen gas and create nitrogen compounds in the soil near the roots of leguminous plants (e.g., soy beans). Nitrates are essential for plants and animals up the food chain to synthesize proteins.

Soil can be solidified into rock, formed from the weathering of rock, and eroded. Therefore, a sustainable supply of *soil* with appropriate characteristics is important for humans. Preventing erosion and nutrient depletion in agricultural soil by rotation of crops, contour plowing, nontill cultivation, and wind barriers, will result in sustainable acreages of productive soil.

What is Soil?

Soil may mean different things to different people, yet all definitions of soil characterize this medium as mostly solid, granular, uncemented, and having interstitial pore spaces.[2] Geologists distinguish between soil and sediment, reserving the latter term to particles in transit or recently deposited, prior to forming rock or developing weathering horizons. Life scientists emphasize the role of soils in growing crops and forming the habitat for soil biota. Engineers (civil) define all Earth materials as soil if they can be excavated by digging. Engineers use soil for construction products and for foundations of engineered structures. Soil, however defined, is an important environmental medium because it can contain nutrients, water, air, soil biota, contaminants, and pathogens. The man-on-the-street calls soil by the unscientific name, dirt, but as students of environmental science, we should not!

Different types of soils have different properties,[3] based on their texture, mineralogy, and structure. Texture is determined by the cumulative weight of different-sized particles, especially the size classes of sand (2.0–0.05 mm), silt (0.05–0.002 mm), and clay (<0.002 mm). Sand particles have the diameter of normal beach sand. Clay particles are so tiny that they cannot be seen individually, but appear in clumps to the naked eye. Silt size is between sand and clay, and feels like flour to the touch. Larger particles—e.g., granules (2–4 mm), pebbles (4–64 mm), cobbles (64–256 mm), and boulders (>256 mm)—can also be part of a soil.

If by weight, the soil is 40 percent sand and 60 percent clay, we call that soil *sandy clay*. A soil having 55 percent clay and 45 percent silt is a *silty clay*. Mixtures of sand, silt, and clay define *loam*, which is the best growing medium. The cohesiveness (stickiness) of clay gives the soil a clayey texture even when the percentage of clay is less than 33.3 percent. Soils that are sandy have the highest permeability, which is the ability to transmit water through the interstitial pore spaces. Clayey soils (high in clay) have the lowest permeability. Clayey soils can hold water and nutrients on clay particle surfaces and between molecular layers of the clay minerals.

Soil mineralogy (or chemistry) also imparts different properties to this medium. Clay minerals like *montmorillonite* have expanding molecular lattices, allowing them to shrink and swell when dried and wetted, respectively. Iron minerals cause rust (iron oxides) to form. *Calcium carbonate* (from limestone) causes soil to be alkaline (high pH). Organic matter added by leaf litter on the surface or root penetration in the subsurface may comprise food for soil biota. Oxidized carbon forms carbon dioxide, which when dissolved in infiltrating water, can form carbonic acid (lower pH).

Soil structure can also affect the properties of this medium. Compacted soil will reduce the volume of pore spaces and, perhaps, the soil's permeability. Roots of plants and weathering can increase the pore space volumes.

Soil has many uses, as drain fields for septic systems, growing media for plants, impermeable layers in landfills, backfill in construction, sandbags for flood control, solid components for drilling muds, and water-well installation, to name just a few. For each use, different mineralogy, structure, texture, and water and air content in soil pore spaces are required. Furthermore, weathering horizons can form in the soil *profile*, providing different properties with depth.

Soil Formation

Soils are formed mostly from the *weathering of* rock in place.[3] The movement of soil constituents in this process is from the top down—that is, from the land surface down to the breakup of rock material. Gravity pulls the soil constituents in infiltrating water to lower levels. An exception to this is in arid soils, where occasionally a rising water table (boundary between water-saturated soil below and unsaturated soil above) brings calcium carbonate upward to precipitate a *hard-pan* or *caliche* layer of calcium carbonate. This precipitate cements soil particles and, thereby, inhibits movement of water through the column of soil. Soil formation involves the interaction of the geo-, hydro-, atmo-, and biosphere, causing the physical and chemical breakdown of rock, the *parent material* of soil.

Physical weathering results from expansion and contraction of exposed rock or soil parent material. This can happen by heating and cooling, by upward heaving of rock from the stress release of the weight of overlying rock and soil, by the breakage and widening of cracks by plant roots or the burrowing of animals, and by the heaving and wedging of ice. These processes loosen the parent material and break it down into smaller pieces with greater surface area. This prepares the broken rock for further development into soil by *chemical weathering* and transport of weathered products to lower horizons.

Chemical weathering results from chemical reactions between parent materials and the combined effects of groundwater, soil moisture, acid or alkaline solvents (from *humus*, plant roots, or acid rain), and oxygen. Unlike physical weathering, where materials maintain their chemical identity, chemical weathering produces new compounds. Carbon dioxide mixes with water to form carbonic acid, and hydrogen and bicarbonate ions. Carbonic acid buffers with calcite to form calcium bicarbonate ions. Carbonic acid combines with sodium feldspar and water to form kaolinite (clay), silicic acid, and sodium and bicarbonate ions. Oxygen combines with iron sulfide (pyrite) and water to form iron oxide (hematite) plus sulfuric acid.

With the physical and chemical changes and the addition of organic matter onto the surface and by plant root penetration, a *soil profile* is formed (**Figure 11-1A**). Different capital letters may be assigned to these profile *horizons*. The letters used here are only one example. Depending on the biome in which the soil forms, the composition and properties of the soil differ.

Common to all different soils is the general characteristics of the profile. The top horizons, called O and A here, are the layers richest in organic matter. The O is the layer of accumulation of organic matter from fallen plant litter and roots of small plants. The A is the layer in part containing partially decomposed organic matter called *humus*. The next lower horizon is the E or zone of eluviation (not shown in Figure 11A), where minerals

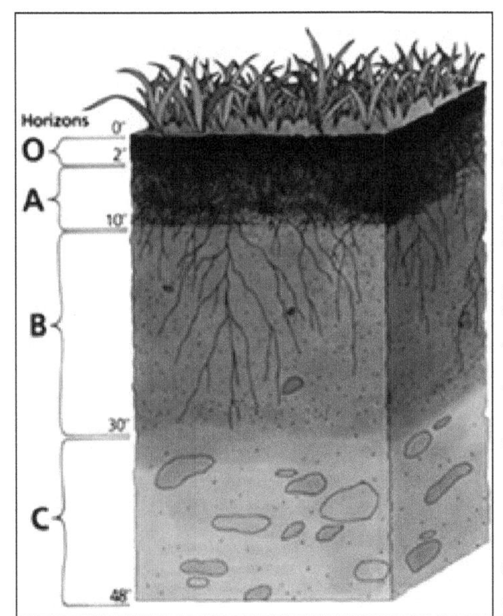

FIGURE 11-1A Soil profile

http://www.nrcs.usda.gov/wps/portal/nrcs/detail/soils/edu/kthru6/?cid=nrcs142p2_054308

FIGURE 11-1B Soil profile

are leached out of the soil and transported to lower horizons. Where do the minerals go that are leached? They are redeposited in the B horizon (or subsoil), immediately below. In moist temperate biomes, the B is rich in clay and insoluble constituents. In dry biomes, the B is rich in calcium carbonate pellets and nodules (*caliche*). In wet biomes, the B is rich in insoluble iron and aluminum oxides. The C horizon can be characterized by broken parent material. Not shown in Figure 11A is the D horizon of unweathered parent material (sometimes called R, for rock, instead of D).

In the field, each of the horizons of a soil profile may not be easily identified because they may not be well developed or visibly blend together. In **Figure 11-1B**, the O (at the grass roots), the A (dark layer below the roots), the E and B (lighter in color below), and the C and D (at the bottom quarter of the photo) can be distinguished.

The surface of a typical landscape usually comprises soils whose parent materials differ depending on the local geology. The characteristics of the soils are further influenced by the climate and type of weathering, the biological activity, the slopes (shallow to steep), and the amount of time that passed during soil profile development. If dynamic geologic changes or periods of severe *erosion* occurred followed by successional changes in the biome, then the soils may undergo successional changes as well.

Soil color can be described based on its hue (spectral shade), its value (lightness or darkness), and its chroma (color saturation). The color of the soil can be used to infer the degree of oxidation (red or brown), low permeability or reducing environment (gray), or content of chemical contaminants (highly discolored).

Soil Erosion

Soil development can be disturbed by the erosive action of high discharges of precipitation or snow melt during short periods of time.[4] This is further exacerbated by the lack of rooted vegetation that cannot take a foothold and hold the soil in place (**Figure 11-2**). Topography can worsen erosion by the absence of stream channels that can contain the

FIGURE 11-2 Eroded soil in sediment-laden stream

runoff, and by the presence of steep slopes. Wind, as well as water, can remove soil and halt its formation.

Disruption of soil development can occur under the influence of gravity flow. This can happen very slowly by creep, faster by rotational block movement, called *slump*, or rapidly with the help of fluids producing mudslides. Soil masses on slopes that are barely in equilibrium could be made unstable by heavy rainstorms, earthquakes, or construction projects that upset the equilibrium. Construction projects that do not consider soil mechanics in their design could destabilize a slope by removing the toe of slopes, increasing the mass of the soil, decreasing the coefficient of friction between the soil and rock beneath, increasing the angle of repose (maximum slope for a stable soil), or allowing water pressure in the soil pores to increase. Any of these processes can trigger a mass movement event, which would disrupt soil development.

Soil Contamination

Natural metals in soil, notably lead and arsenic, can cause health problems (neurologic disorders and cancer). *Manganese* leached from soils and percolating into the groundwater is associated with psychiatric disorders and mental diseases. In addition, dumping wastes onto the land over permeable soils (a common practice prior to strict waste-management regulations) contaminates groundwater aquifers and results in adverse health effects for the nearby populations **(Figure 11-3)**.

FIGURE 11-3 Stained soil from mining (now abandoned)

References

1. Enger, E. D. and B. F. Smith. 2008. *Environmental Sciences: A Study of Interrelationships,* 12th ed. New York: McGraw-Hill.
2. Harpstead, M. I. 2001. *Soil Science Simplified.* Ames: Iowa State University Press.
3. Gerrard, J. 2000. *Fundamentals of Soils.* New York: Routledge, 230 pp.
4. Miller, D. T. and D. T. Gardiner. 2001. *Soils in Our Environment.* Upper Saddle River, NJ: Prentice Hall.

12 Crustal Rocks and Sediments

We sometimes lose sight of the fact that raw materials that we need for our daily lives are derived from Nature, not from the hardware or building supplies stores. Products for buildings, dams, roads, tunnels, manufactured items (including chemicals), and coal come from *crustal rocks* or *sediments*.[1] We also tend to forget that crustal rocks and sediments are conduits for the movement and collection of water (groundwater aquifers and surface drainage basins and impoundments) and other fluids (oil and natural gas), which are also vital to our existence. Finally, weathered rock and sediments are the parent materials for agricultural soil, needed to grow our crops.

We need to maintain a sustainable supply of economically exploitable materials derived from crustal rocks and sediments. This may require recycling products derived from already extracted sources. Also, we must maintain these supplies without causing pollution from the excavation and processing of the extracted materials.

Crustal Rocks

A sustainable supply of rock products is necessary for manufacturing products and for materials used in construction and earthworks. Rocks are naturally formed from solidification of molten *magma* (*igneous*), from deposition from fluid media (water and air) (clastic *sedimentary*), from precipitation from aqueous solutions (*evaporites*), and changed by heat and pressure (*metamorphic*). Evaporites are also considered sedimentary, but not clastic (broken particles).

Although *granite*s are only one of many *igneous rock*s, the layman term *granite* is applied to all igneous rocks that contain visible mineral crystals. It is used in building stone, gravestones, kitchen countertops, and flooring for its strength and resistance to weathering in temperate climates. Igneous rocks whose crystals are not visible or partially visible were derived from molten rock that was extruded onto the surface as *lava*, *pumice*, *volcanic ash*, or *obsidian* (volcanic glass). Pumice **(Figure 12-1)** has been used as an abrasive or a skin defoliant. Volcanic ash (*bentonite*, whose clays are expandable when wet) is used in drilling and installation of water wells. *Obsidian (silica)* is hard and breaks into jagged pieces, which made its use by early humans as projectile points and tools; now archaeologists use these fragments to study early human cultures.

Clastic sedimentary rocks include *sandstone* **(Figure 12-2)**, *limestone*, and *shale*. They, too, are used as building materials, including building stone, paving stone, and road and railroad tie ballast. In buried layered structures that have been faulted and folded, the

© Alan Jacobs

FIGURE 12-1 Pumice deposit on Lipari, Eolian Islands, Italy

FIGURE 12-2 Sandstone outcrop

rocks have formed traps for the collection of petroleum and natural gas. A rock similar to sandstone called *conglomerate* (rounded clasts) or *breccia* (angular-shaped clasts) has been used as grinding-stone wheels in grist mills. Some *limestone* and shale contain fossils which are used to study past life on Earth. Permeable clastic rocks, (e.g. some sandstones) in place can be productive aquifers.

Evaporites, chemically precipitated sedimentary rocks, contain a concentrated supply of useful products. Precipitation and crystallization of each product takes place at certain temperatures and pressures in water, and the process concentrates these products as the water evaporates (thus the name *evaporites*). Evaporites include table or *sea salt* (sodium chloride, a food additive), gypsum (calcium sulfate, used in wallboard), and other compounds with many uses including borax (sodium borate), epsom salts (magnesium sulfate), and trona (source of sodium carbonate). Sea salt is crystallized by the evaporation of modern-day ocean water, whereas much table salt is quarried from ancient salt beds that were formed and buried for millions of years; the evaporitic process is the same as for sea salt and the chemical makeup and nutritive value is similar.

Limestone can also be an evaporite. As such, it is usually pure enough in calcium carbonate to be used where that compound is needed in a pure form. Crushed limestone is used in making cement, in steel manufacture, and, in powdered form is used in antacids for digestive distress. Limestone-derived waste from blast furnaces at steel mills produces a glassy material called *slag*. Slag is also used as a building material and ballast, and is mistakenly identified as a meteorite by layman. Limestone blocks (either evaporitic or clastic) are also used as building stone (e.g., Empire State Building, Pentagon, National Cathedral) and gravestones, especially in the arid regions where low rainfall does not chemically erode it over long periods of time.

Metamorphism alters other rocks in their mineral composition and orientations, without melting the altered rock. This usually results from deep burial or contact with molten rock or hot aqueous solutions (*hydrothermal*). Metamorphic rocks mimic some igneous rocks, but the former often show a linearity of layering from the metamorphic process. Uses for metamorphic rocks are similar to those for igneous rocks. The most famous of metamorphic rocks is marble, which is metamorphosed limestone that is used for building materials and sculptures.

FIGURE 12-3 Sandbags holding back floodwaters

SEDIMENTS

In a way, sediments could be described as clastic sedimentary rocks that have not yet been solidified or cemented. Sediment includes *clay*, silt, sand, gravel, cobbles, and boulders in size order from finer to coarser. Accumulations of these sedimentary particles and their mixtures are a necessary resource for society. Clay is used to make bricks, ceramics, and liners for landfills (to prevent leachate from seeping into groundwater aquifers). Sand is used for flood control (sand bags) **(Figure 12-3)**, and glass manufacture. Mixed sediments are used as construction backfill and in earthen dams. Gravel is used as aggregate for concrete. Gravel is also used as road metal, railroad tie ballast, and in drain fields in septic systems. Permeable sediments (sand and gravel) in place in thick lenses or filling buried glacial bedrock valleys can be productive aquifers.

PRODUCTS WORTH MINING OR QUARRYING

There is gold, silver, uranium, clay, and other potentially useful metals and nonmetals in most rocks and sediments; however, they may be in trace quantities that are too small to be economically exploitable. If the cost of extracting (separating them from the rock or sediment) and processing is more than the value of the material sought after, the material is not exploited.

Ancient geologic processes, however, have concentrated some of these valuable materials. Economically viable materials can be found in so-called veins, fracture fillings, metamorphic contact zones, gravel terraces, pockets of sediment sorted by streams or ocean waves, lake clays, or rock formations containing high concentrations of lime, gypsum, sand, salt, sulfur, and other deposits. Knowledge of geology is necessary to discover their locations.

Micronutrients

Although exposure to high concentrations of many metals can be toxic, small amounts of certain metals are vital to health. These are the *micronutrients* and include copper,

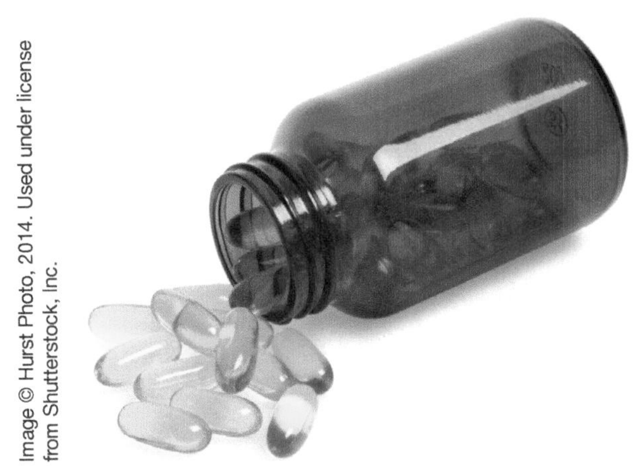

FIGURE 12-4 Mineral supplement pills

iron, and zinc, necessary for growth and development. It is advisable to obtain these micronutrients in a balanced diet, or, if necessary, in so-called *mineral supplements* **(Figure 12-4)** under the advice of a physician and sufficient to meet recommended daily amounts (RDAs).

A Sustainable Supply of Crustal Materials, without Pollution

In the processing of these materials, undesirable waste products are produced—potentially causing pollution of air, water, or soil. A wider (but perhaps incorrect) use of the term *minerals* refers to nonfuel-related products of petroleum and coal, including plastics, solvents, food dyes, and other chemicals. A sustainable supply of minerals is necessary to manufacture most useful things we see around us. Whatever their origin on Earth, they should be processed with no or a minimum of waste products that might cause pollution.

References

1. Riley, C. M. 1959. *Our Mineral Resources*. New York: Wiley.

13 Food

Image © athurstock, 2014. Used under license from Shutterstock, Inc.

A Sustainable Food Supply

As consumers in the Animal Kingdom, humans must eat to survive. The quantity and quality of the food we eat determines the sustainability of our society, our health and safety.

For a healthy and safe diet, we must eat enough, not overeat, eat a balanced diet including macro- and micronutrients, and beware of toxins and pathogens in the food that might harm us. We must be able to afford to buy or grow the accessible food products, and then prepare and consume them properly and regularly.

For some of us, food intake may be more complicated. We might have dietary restrictions resulting from medical conditions (food allergies, diabetes, obesity), from ethical or religious reasons (no meat, no pork or shellfish, free-range), from environmental concerns (locally grown, low on the food pyramid, organic), or from the inability to afford food for an adequate and healthy diet. Many of us just binge on what is tasty and easy to obtain, without regard to nutritive values (fast foods, preprepared foods, snack foods, sweets). Although the food industries provide a wide variety of foods, it is up to us to choose wisely.

Agriculture (Origins)

Unlike other consumers in the Animal Kingdom, most humans do not rely on hunting and gathering for their food supplies. We once did. Of course, our population then was of the order of thousands, not the current 7 billion mouths to feed.

About 5,000 years ago, we started growing our own food, a technique called *agriculture*[1] **(Figure 13-1)**. Agriculture refers to methods of growing or raising food for others (and ourselves) to eat. With a portion of the population occupied with agriculture providing food for everyone, other people could do other things, such as making tools, making clothing, preparing food, and manufacturing goods. This allowed humans to multiply way beyond what was possible under hunting and gathering. Agriculture, then, was important to people in all other professions.

Consequently, human civilizations grew up where soils were ideal for growing plants. Such soils were found on floodplains of major rivers. The floodplains were the sites of deposition of mineral-rich silts when the channels would overflow their banks during the rainy seasons.

Three features of the floodplains helped to develop and maintain a rich soil as a growing medium. As the waters overflowed onto the floodplains laden with silt, the velocity of the water was significantly reduced preventing the silt from remaining in suspension. Most of the silt, therefore, was deposited at the edges of the channels. This built up what are called *natural levees* (a natural alternative to the artificial levees constructed nowadays by the U.S. Army Corps of Engineers). The levees reduced river waters from overflowing their banks in subsequent floods. Consequently, the soils were eventually less erodible.

A second soil development and protection feature was the natural growth of plants on the floodplains. Roots held the soils in place and the biomass from the plants added needed organic matter to the upper soil horizons. The rivers were also a magnet for animals to gather for water, preying on aquatic life, and preying on other animals coming to the floodplains. Their wastes also added organic matter to the developing soils.

A third feature helped to develop and sustain a rich soil. When rivers cut deeper channels (during periods of land uplift and sea level fall), the parts of the floodplains that were preserved became terraces that were too high to be flooded by the channels that were now at lower elevations. These terraces also became good areas for agriculture.

FIGURE 13-1 Floodplains were suited for agriculture

Thus, centers of human civilization flourished along floodplains of the Tigris–Euphrates and Nile river systems. Hunting and gathering eventually became relegated to sporting and recreation activities.

Agricultural Development

Primitive agriculture spread from the floodplains to other areas where, depending on the biomes, the natural environment had to be changed to suit agricultural requirements. Natural vegetation tended to have *biodiversity*, numerous species of plants that change to conform to the natural biome of the surrounding area.[2] Agriculture, on the other hand, required the development of fields or rows of one specific species of plants to be efficiently planted, cultivated, and harvested. Furthermore, erratic weather was not ideal for watering the crops; so irrigation had to be added. In addition, soil nutrients that became depleted had to be replenished by added fertilizer (initially the mixing of manure into the soils). Finally, pests (mainly weeds and insects) had to be kept at bay.

Certain biomes were better suited to agriculture. Aside from former floodplains (and some terraces), grasslands were better than forests (more nutrients in the soils), temperate biomes were better than boreal (longer growing seasons), and moist climates were better than deserts (more available water). Although the tropical rain forest might appear to have good growing conditions, the soils of these forests and temperate forests have poor soils—most of the nutrients are contained in the bodies of the trees and other vegetation.

Therefore, the clearing of forests provides the worst possible soils for agriculture, other things being equal, and has other negative consequences. *Clear-cutting*[3] removes all trees in an area without a chance for the forest to regenerate itself and produces poor farmland. *Slash and burn* involves cutting the trees and setting them afire; then the burned vegetation is plowed into the soil. This frees some of the organic matter from the trees but adds to the soil only a small amount of nutrients for one or two growing seasons at most. The best possible conversion of natural biomes to farmland involves the use of grasslands, which has soils rich in organic matter and other nutrients. Transforming forests into farmland also removes many environmental services that forests provide, for example, wildlife habitats, flood control, erosion prevention, and recreational areas. If farmland is needed, it should come from conversion of temperate grasslands where rainfall is adequate.

Agricultural Trends

Early forms of agriculture involved small plots, *polyculture*[4] (mixed crops), intense manual labor and the use of beasts of burden (especially for plowing the soil). This kind of activity did not use mechanization (motorized machinery). Modern trends, especially in sparsely populated areas, have shifted to *monoculture* (one crop over large acreage) using motorized machinery. Mechanized agriculture requires machinery, fuel to run the machines, monoculture for efficient production, and using a seed source that grows produce having uniform characteristics.

An agricultural program, confusingly named the *Green Revolution*,[5] started in the mid-twentieth century. It was named "green" not because it was environmental friendly, but because food production was greatly increased worldwide. This was accomplished by the introduction of new plant varieties, by farming methods that prevented erosion, by stepped-up usage of fertilizers and pesticides, and by increasing the amounts of irrigation waters. One of the downsides of this revolution was that overly generous amounts of pesticides and fertilizers were applied by farmers to their fields. The surplus was not completely retained in the soil. Instead, it washed off the surface in precipitation runoff, which emptied into lakes, ponds, rivers, and wetlands and infiltrated into groundwater aquifers. The production of these chemicals also entered environmental media by improper waste handling and spills.

There are business advantages to growing uniform produce. Buyers can rely on a standard of quality, price, and characteristics. Fast-food vendors want potatoes that make uniform sizes and tastes in French fries. Manufacturers of cattle feed want the corn and soybeans to have uniform nutritional values. Seasonal farm laborers could be hired to repetitively pick the same vegetable in an efficient manner. Farmers could buy hybrid seeds having the same characteristics.

There are disadvantages of planting the same species (or variety) of seed, mainly the possibility that a blight affecting that variety could wipe out the entire crop. This was one of the problems leading to the *Great Famine of 1845–51* in Europe,[6] where the potato blight (caused by an oomycete called *Phytophthora infestans*) damaged the tubers. Ireland, dependent on one variety of potato for a major part of the diet, was most severely affected, resulting in millions dying or emigrating to other countries (notably the United States). Such monocultural planting recently also affected the corn plants in the US Midwest and grape vines in California during blights that made the crops vulnerable. Insect pests also targeting specific crops have an easier time spreading over large areas.

Mechanization in agriculture also results in large tracts being plowed long before planting occurs, leaving fields subject to erosion during the period the plowed land remains exposed **(Figure 13-2)**. This trend also favors large corporate agribusiness at the expense of small family-run farms.

FIGURE 13-2 Mechanization on large tracts of land

Modern Problems with Agriculture

Greater use of chemical fertilizers results in *eutrophication* of nearby surface waters producing algal blooms. Sunshine cannot penetrate to lower layers of thick algal mats, resulting in the deatlh of algae. Dead algae are consumed by decomposing organisms. This uses up the DO in the water. The resulting *hypoxia* (low oxygen levels) in these waters kills aquatic life below the mat.

Even though fertilizers provide mineral nutrients, they do not replace vital organic matter in the soil. Greater use of chemical pesticides pollutes water supplies. Finally, there is recurring erosion and dependency on irrigation.[7]

Solutions to the Problems of Agriculture

Environmentalists suggest returning to less disruptive forms of agriculture, although this does not necessarily result in the massive quantity of food needed for the existing and future populations. Without severe population control, which is unlikely, some of the traditional practices in farming (small farms, fertilizing with manure, organic farming, polyculture, *nontill farming*) could not keep up with food demand.

Nevertheless, production requirements could be met with some proposed environmentally approved agricultural methods[8]: *contour plowing*, *rotation of crops*, wind barriers, grassy runoff tracts, efficient irrigation targeting the roots of plants, and *integrated pest management* (natural and chemical means combined). Not approved by environmentalists because of untested food safety potential is the use of *genetic engineering* (banned by many western European countries) to increase yield, resistance to pests, and modification of produce characteristics (taste, color, size, etc.).

Meat, Fish, and Seafood

Agriculture includes not only the production of crops (plants) for food supply. It also includes the production of animal products, including meat and seafood.[9,10] Agriculture is also involved with nonfood production in the form of textile products from fibers derived from both plants and animals. Aquiculture, akin to agriculture, also produces animal products from aquatic environments.

Eating animals is biologically possible for humans. We are omnivores; we can digest both animal and plant tissues, within limits. We cannot, however, digest animal bones or woody plant tissues. Nevertheless, although digestible, certain foods are not eaten because some consumers have self-imposed food limitations.[11]

Vegetarians, humans who choose to not eat animal tissue, do so for ethical or environmental reasons. Killing animals that fear death or experience pain is believed by those who have eliminated meat from their diets to be immoral for humans. Vegans go one step further. They will not eat products derived from animals, such as milk products or eggs. Then there are others who feel that plants should not be killed either. Fruits produced by plants to spread their seeds are designed to be eaten. Wildlife eats fruits (all fruits have seeds) and then they defecate feces with seeds to start the growth of new plants. Eating fruit does not kill the plant itself. People who will eat only fruits, are called *fruitarians*. Then, there are others who call themselves "*breatharians*," those who claim not to eat or drink. Although there are claims of survival from those who do not eat or drink anything, such claims are scientifically unsubstantiated.

Plants collect energy from the Sun, and are eaten by *herbivores* and *omnivores*. Omnivores and *carnivores* eat animals that have eaten plants or other animals. At each transfer of energy up the *food chain* (from prey to predator), energy is lost. The transference is about 10% efficient with each transfer. Therefore, it is more energy efficient to eat lower on the food chain—that is eating plants rather than animals.

Some people will eat meat if they have the producers' assurances that the animals were slaughtered humanely and kept alive under humane and environmentally safe conditions. Caged poultry in cramped quarters is not a humane practice. The eviscerating (gutting) of a hog while alive (*pig-sticking*) is not a humane practice. The collection of large numbers of cattle on a *feedlot* to fatten them up before slaughter produces huge masses of waste-laden runoff. Such feedlots can contaminate nearby water supplies. Hunting is decried by some people when practiced only as a recreational sport. Animals killed during hunting should be used for food.

Fish and other *seafood* is another source of meat. We derive this source of food commercially from harvesting fish from the oceans and freshwater lakes. We also use *aquiculture*, a process like farming on land, which cultivates aquatic life in freshwater ponds or saltwater cribs in ocean bays.

Several environmental issues result from fishing or aquaculture.[12] Overfishing can reduce or destroy species of fish, reducing the population of various aquatic species to levels that threaten survival of the species. *Bykill* results when nontargeted sea creatures get caught in nets intended for targeted fish (e.g., dolphins get caught in tuna nets). Ingesting of toxic substances that are disposed directly into or drained into water bodies can kill aquatic life. Such toxic substances can *bioaccumulate* through the food chain and *bioconcentrate* in the fatty tissues of aquatic organisms.[13]

A USEPA study on mercury content in fish and other seafood found unsafe levels of mercury in shark, swordfish, and king mackerel and low levels (but present nonetheless) in shrimp, canned light tuna, salmon, pollock, and catfish. The fish and sea creatures that are large and have lived longer have had a greater potential to bioaccumulate more mercury (and other contaminants) in their tissues than other smaller creatures with shorter life spans. Additional information on mercury in fish and shellfish[14] may be found at the *Food and Drug Administration (FDA)* website (http://www.fda.gov/Food/Food*Safety*/Product-specificInformation/Seafood/FoodbornePathogensContaminants/Methylmercury/ucm115644.htm).

Should you eat only small fish and seafood? You may reduce your consumption of contaminants if you eat the flesh of the fish and not down little fishes whole (as with smelts or sardines), which include organs like the liver that concentrates toxins.

Food Safety

In the United States in one year, more than 75 million people contract food-borne illnesses, resulting in about 5,000 deaths.[15] These illnesses can be prevented through proper handling of foods. Prevention requires public health inspection, training of food handlers, and consumer education.

Our food is not pure animal or plant tissue. We derive nutrients and fiber with each bite, but there are other things in it—traces of pesticides, feces, spoilage retardants, growth hormones, precipitants from air pollution, pathogens, parasites, drug residues, coloring agents, and allergens. Even organically raised produce can have unwanted contaminants, especially insect feces and eggs. Where do these things come from? We are not the only potential consumers of animal and plant tissue that we call our food. A delectable food product is also on the menu of other organisms, large and small. Furthermore, plants manufacture their own natural pesticides to prevent themselves from being eaten.

We can reduce our intake of food contaminants by washing the food, cooking the food, choosing organically-grown food, avoiding the eating of organs that collect contaminants in relatively high concentrations (i.e., liver), and avoiding prepared foods that have a considerable amount of additives to prevent spoilage and improve the taste and color.[16] We can avoid foods with added artificial flavors and colors. We can grow our own food—getting back to the traditional agriculture of our ancestors. Our careers and busy family lives, however, might prevent us from spending too much time on gathering and preparing what we eat.

Our immune system can combat the contaminants in foods up to a point. Humans are born with certain inalienable immunities. Among these are resistances[17,18] to many pathogens, chemicals, and parasites. Humans who could not deal with the most common contaminants did not survive, and did not pass on such genetic traits to future generations. Our ancestors survived with certain immunities, attesting to our existence today. How do we protect ourselves? Using metabolic catalysts, called *enzymes*, our bodies can change harmful contaminants into less harmful contaminants. We can also protect ourselves from these contaminants by changing the contaminants into chemicals that can be more easily excreted before they can do harm. Many generations of evolution have "trained" our bodies to react this way. Unfortunately, we do not have this "training" to combat new chemicals, to date more than 60,000 of them in commercial use. The following examples show that some chemicals can be beneficially metabolized, others metabolized into more harmful substances, and others dealt with successfully only at certain concentrations:

Can we tell if food is not suitable to eat[19]? Unfortunately, the telltale signs of unsafe food are not foolproof. Bloated canned goods, mold on produce or dairy products, evidence of feces, or rancid meats are obvious signs of foods to avoid. Yet, food that looks and smells okay still may not be safe. *Public health inspections* of food services (restaurants) and establishments (groceries) look for telltale signs of unsafe food, but they also look for the way the food is kept. Unsafe storage and handling can lead to the growth of microorganisms. Public health inspections look at the temperatures of stored foods, cleanliness of preparation surfaces, food acidity, and possibilities of cross contamination of both foods and preparation utensils.

What makes food unsafe to eat? Microorganisms such as bacteria, viruses, parasites, and fungi can grow in food and in our bodies and produce toxins that can damage cells in our bodies. Microorganisms can be introduced to unprotected food by insects (**Figure 13-3**) and rodents **(Figure 13-4)**. The dangerous environments of food include temperatures between 41°F and 140°F for more than 4 hours and a pH of 4.6 to 7.5. Consequently, one must keep the food (carbohydrates and proteins) in refrigerators or freezers, cook it to well-done temperatures, or keep it in an acidic (pickling) environment. In addition to temperature and acidity, moisture and oxygen availability also affect food safety.

Moisture can encourage bacteria to grow, especially water, which has a potential to encourage chemical reactions. This property is called *water activity*,[20] represented by the symbol a_w. The property a_w in foods is similar to the property of relative humidity in air. Water vapor in air will flow from humid air (higher energy) to less humid air (lower energy). Water vapor will also be absorbed by food, which has a lower energy level than the surrounding humid air. The worst a_w for a safe food environment has an a_w of 1.0 (equal to distilled water). The best a_w for a safe food environment is a value of less than 0.5, in which no bacteria or mold can grow.

FIGURE 13-3 Unprotected food infested by a housefly

Water activity regulations are applied to foods with an a_w value greater than 0.85. For example, liverwurst, cheese spreads, and caviar have values of a_w that are higher than 0.9. Foods with low a_w values have a longer shelf life. Depending on the species of bacteria, the a_w can restrict that species from growth. Luckily, the most virulent bacteria, strains of *Clostridium botulinum* can only grow with an a_w higher than 0.97 and most other bacteria (not as virulent) grow with an a_w of greater than 0.9. The a_w is reduced in foods that contain curing agents and are kept at lower temperatures, vindicating the need for refrigeration, or at best curing the food with salt or sugar or dehydrating them.

Oxygen can also lead to food spoilage. Oxygen, vital to us, is also vital to microorganisms. They can grow faster with available oxygen. Canning removes most of the oxygen in the food and seals additional supplies of oxygen from nurturing the microorganisms or their eggs or spores that had not been killed by the heat applied during the canning process. Oxygen also reacts with the food, either with the help of enzymes or with normal chemical oxidation. The browning of peeled fruits is an example of oxidation with the help of enzymes.

Consequently, the food (F), acidity (A), temperature (T), time (T), oxygen (O), and moisture (M) are the parameters that affect food safety with respect to bacterial growth.

FIGURE 13-4 Unprotected food infested by a mouse

The first letters of these words in this order spell out an acronym for remembering these parameters: *FAT TOM*.[21]

Ignoring the ways food can be unsafe can lead to adverse health effects. *Food-borne illnesses* can lead to infection due to bacteria, viruses, *parasites*, and fungi. Again, we have some natural immunity to fight the infections, but this depends on our resistance, the type of microorganism, and the concentration of the microorganisms ingested.

Bacteria are a major concern of food- and waterborne illnesses. These include *Escherichia coli (E. coli)*, *Salmonella*, *Shigella*, *Listeria*, and *Bacillus botulinus*, among others. They can produce mild to severe cases of gastroenteritis, and even death. They can proliferate in the food and in our bodies.

Viruses ingested through food are also a concern. Such viruses include *hepatitis A*, *Norwalk virus*, and *rotavirus*. The viruses replicate in human bodies (in tissue cells), but not prior to ingestion. They cause liver damage (hepatitis A) and gastroenteritis.

Parasites are organisms that live within us and cause us harm. Their bodies or their eggs or cysts take up residence in the animals we eat and continue their life cycle in us. Such parasites include *Trichinella spiralis* (larvae of roundworms that cause trichinosis), *Giardia*, and intestinal *Cryptosporidia*.

Fungi, including molds, yeasts, and mushrooms, grow on foods and cause spoilage and illnesses from the toxins they produce.

References

1. Cohen, M. N. *The Food Crisis in Prehistory: Overpopulation and The Origins of Agriculture*. New Haven, CT: Yale University Press.
2. Pollan, M. *The Botany of Desire: A Plant's Eye View of the World*. New York: Random House.
3. Gorte, R. W. *Clearcutting in the National Forests: Background and Overview*. Washington, DC: CRS.
4. Connor, D. J. *Crop Ecology: Productivity and Management in Agricultural Systems*. New York: Cambridge University Press.
5. Dahlberg, K. A. *Beyond the Green Revolution: The Ecology and Politics of Global Agricultural Development*. New York: Plenum Press.
6. Percival, J. *The Great Famine: Ireland's Potato Famine, 1845-51*. New York: Penguin.
7. Burley, R. *Agribusiness: Problems and Solutions* (video recording). Princeton, NJ: Films for the Humanities & Sciences.
8. Beeman, R. S. and J. A. Pritchard. *A Green and Permanent Land: Ecology and Agriculture in the Twentieth Century*. Lawrence: University Press of Kansas.
9. Hutchinson, L. *Ecological Aquaculture: A Sustainable Solution*. East Meon, England: Hyden House.
10. Field, T. G. *Scientific Farm Animal Production: An Introduction to Animal Science*. Boston: Prentice Hall.
11. Goodman, D. *Alternative Food Networks: Knowledge, Practice, and Politics*. New York: Routledge.
12. Tyus, H. M. *Ecology and Conservation of Fishes*. Boca Raton, FL: CRC Press.
13. Beek, B. et al., eds. *Bioaccumulation: New Aspects and Developments*. New York: Springer-Verlag.
14. U.S. Food and Drug Administration. Mercury Levels in Commercial Fish and Shellfish (1990–2010). http://www.fda.gov/Food/FoodSafety/Product-specificInformation/Seafood/ FoodbornePathogensContaminants/Methylmercury/ucm115644.htm

15. Heldman, D. R., M. B. Wheeler, and D. G. Hoover, eds. *Encyclopedia of Biotechnology in Agriculture and Food.* Boca Raton, FL: CRC Press.
16. Rehe, S. *Preventing Foodborne Illness: A Guide to Safe Food Handling.* Washington, DC: U.S. Department of Agriculture, Food Safety and Inspection Service.
17. James, J. M., W. Burks, and P. A. Eigenmann, eds. *Food Allergy.* New York: Elsevier.
18. Rich R. R. et al., eds. *Clinical Immunology: Principles and Practice.* Philadelphia: Mosby.
19. Merino, N., ed. *Food Safety.* Farmington Hills, MI: Greenhaven Press.
20. Barbosa-Canovas, G. V. et al., eds. *Water Activity in Foods: Fundamentals and Applications.* Ames, IA: Blackwell.
21. Matthews, D. D., ed. *Food Safety Sourcebook.* Detroit, MI: Omnigraphics.

14 Technological Infrastructure

Our Technology Provides Safeguards

If we had to fend for ourselves in the natural environments, we would have the same limitations for survival as wildlife. We would be the prey of many creatures most of us see now only in zoos, on safari, and in travel films. In the wild, we would be at the mercy of severe weather conditions with only our bodily hair and primitive natural shelters to protect us. We would lack the knowledge of impending climate changes and natural disasters, and how to prepare for them. We would be hunters and gatherers ourselves, without a food supply from agriculture. We would succumb to many diseases and infections without access to medical help. Banding together, we might, like wolves, bees, herds, etc., derive some protection in numbers and specialized niches. But this would be a wild existence, without the benefits of our technological development.

FIGURE 14-1 Comforts of home and technology

Survival of the Most Comfortable

We, however, developed a civilization that keeps improving on the way we cope with survival.

After thousands of years, we have learned to separate ourselves from the natural environment with a technological infrastructure.[1] Indoors, the temperature is regulated to our comfort zone, warmed in the winter and cooled in the summer. Even the humidity can be regulated to suit our preferences. Indoor plumbing brings fresh water into our homes and conveys wastewater out. We have artificial lighting; furniture for relaxing, eating, and sleeping; appliances to help with cooking, cleaning, and washing (dishes and clothes); and telephones, televisions, radios, computers, and other electronic devices for communication and entertainment. We have security devices like locks, alarms, and surveillance cameras to keep out and identify intruders, screens on the windows, pesticides to keep out insects and rodents, and areas for storage, study, and entertaining visitors and ourselves. The higher our economic status (or the more we are in financial debt), the more of these manufactured comforts we have—or, at least, buy and intend to use. The lower our economic status, the fewer of these comforts we can afford. We might call this survival of the most comfortable **(Figure 14-1)**.

Evolution of Technology

Technologically, all living creatures have evolved to mimic the natural evolutionary development of life on Earth. Life has evolved so that the organism can either regulate its body temperature (warm-blooded, or *endothermic*) or move to areas where the ambient temperature suits the body's needs (cold-blooded, or *ectothermic*). Creatures that need light are active during the day (*diurnal*); creatures for which hunting is better by night are active in the moonlight (*nocturnal*). In the deep sea, where sunlight does not penetrate (below 600 m), some animals have developed *bioluminescence* to glow in the dark. Creatures find shelter in trees, burrows in the ground, nests, caves, and other forms of housing they create or find for themselves. They communicate with others through vocal (roar of the lion), sonar-like emissions (bats and whales), patterns of movement (insects), courtship rituals (many animals), chemical odor emissions (many animals), *mimicry* (plants "looking" for pollinators), and protective repellents (for example, thorns or poisonous tissue in mushrooms and poison ivy).

When we humans venture out of our rooms to go to work, school, play, vacation spots, health facilities, and shopping malls, technology follows us with the aid of manufactured infrastructure and other indoor and outdoor environments for transportation, communication, and other activities. We have structured our lives to survive in relative comfort with the help of technology.

Yet, when we think about the beautiful forests, mountains, and seacoasts, we imagine unspoiled landscapes of our dreams. Biologists call these areas *biomes*, which also include *grasslands* (including *savannas*); *tundra*; forests (boreal (northern), temperate, or rainy); *aquatic environments*; *deserts*; and ice-covered land. Each biome is defined on the basis of temperature range, available moisture, soil types, and assemblages of living organisms that thrive in this environment. Books, magazines, television programs, and video travelogues beckon us out of our rooms and into these natural biomes as a way to escape from our protected, technologically manufactured environments and to enter the pristine world of Nature, welcoming and safe.

What is wrong with this picture? Two things are hidden. The indoor, manufactured and technologically improved environments are not as safe as advertised—but then neither is Nature as safe as described in the guidebooks and brochures.

Do not get too comfortable in your high-tech room until you check things out. Is mold growing behind the paneling? Is the forced air intake of your heating and cooling system coming from outdoor areas that are feeding polluted air into the house (from garbage-filled dumpsters, automobile exhausts, garden and cleaning chemicals, or even mold growing in damp environments)? Are your artificial fiber carpet and its adhesive giving off vapors into your breathing zone? Are your heating and plumbing systems properly venting gases away from the house. Are particulates in breathing zone air being filtered prior to entering our respiratory system and making their way to our lungs and bloodstream? Are all sharp objects and cleaning and pesticide chemicals kept away from young children? Is your water supply monitored regularly to ensure there is no contamination from sewage? Are you living over abandoned mines, near waste dumps, near high-voltage electric transmission lines, downwind from smoke stacks, or over bedrock that emits radon gas?

Okay, now that you are leery of the indoor environment, let's venture outside. Imagine the following:

An interplanetary vehicle lands on a body in space, and the captain wants to explore the environment outside the ship **(Figure 14-2)**. First, she asks the science crew if it is safe to venture forth. The initial environmental parameters measured are oxygen concentration, temperature, and radiation levels. Next, the crew searches for signs of alien life forms.

What if this spaceship had landed on a planet called Earth? The following natural dangers might be reported to the captain:

"Captain, our sensors indicate:

1. Natural toxins (poisons) and pathogens (disease-causing organisms) are detected in potential food and water sources.
2. There is the possibility of infections and parasite attacks from organisms in the soil and biota.

FIGURE 14-2 Prepared for a hostile environment

3. There is the risk of falls, fires, and bites from wild animals, including insects.
4. Certain areas might experience severe weather phenomena, like tornados and hurricanes. Other areas might be subject to severe storms producing lightning, hail, snow and ice, damaging winds, and floods from torrential rains and snowmelt.
5. Certain zones could experience earthquakes, crustal shifts, tsunami, landslides, and ground subsidence.
6. High radiation levels could be found near uranium deposits, especially radon gas, and even in most areas during the day from UV radiation in normal sunshine.
7. Grasslands and forests during dry seasons could experience lightning or spontaneous combustion induced fires (forest and range).
8. Areas near active and potentially active volcanoes can produce molten rock streams, hot choking ash falls, poisonous gases, and eruption-induced earthquakes."

In fact, the environment of this planet is so fraught with hazards that the chief science officer might recommend that the outside crew don spacesuits with tanks of supplied air, in spite of the fact that many Earthlings can be seen wandering around in shorts, T-shirts, and flip-flops. The assessment continues:

"Captain, there is evidence of a technologically advanced human society on the planet. As a result, we see that there are additional environmental dangers, namely:

1. Air and water pollution from factories and power plants
2. Abandoned waste sites and illegal dumps that have not been sufficiently cleaned up
3. Wars and terrorist attacks from human groups that do not agree with each other
4. Transportation accidents, crime, poor sanitation systems, and unequal access to medical care
5. Overpopulation, poverty, and dangerous workplace conditions
6. Siting of nuclear power plants and dams in unstable crustal areas.
7. Dwindling energy supplies and lack of foresight to better develop and rely on nonpolluting, sustainable sources of energy"

The captain might conclude that this planet is too hostile to explore. She might also note that humans have added many hazards to those caused by Nature. She might well ask: Why was not the public educated in environmental science so that they could have removed or lessened the impacts of the hazards that they themselves have caused?

We should all be wondering the same thing! But as you are here taking this course to become aware of environmental problems and how to solve them, we will acknowledge that we may be "preaching to the choir." But there is still the challenge of reaching those who have not considered environmental science as a basis for protecting the health and safety of society. All we can say is—spread the word! To do so, you must be able to explain some basic scientific principles covered in the first chapters.

References

1. Jacobs, A.M. and D. Porter. 2012. *Environmental Science – Health Impacts.* Dubuque, IA: Kendall Hunt Publishing.

UNIT 4

Energy Resources

Nonrenewable (Nonsustainable) Energy Resources

Generating and Using Electricity

One of our energy resources involves electricity.[1] We generate *electrical energy* generally by burning fossil fuel or using nuclear fission to create steam from boiling water. We direct that steam against the blades of a turbine which turns a turbine wheel. The wheel contains coils of electrical wire that, when turned in a magnetic field, makes the electrons flow in the wire. This flow of electrons is electricity. Electricity in turn can energize electrical devices, e.g., lamps, heaters, motors.

Electrical energy can be stored in batteries for later use. Until efficient and lighter batteries were produced, an electrical cord had to deliver the flow of electrons to those devices or lightweight batteries could be used only in low-energy mobile or portable devices, e.g., flashlights, radios, small toys. The weight of batteries limited the use of electricity in transportation vehicles. We recently, however, started making automobiles using battery assist (hybrids) and total electric cars. The cars that do not need fuel fluids have limited performance, i.e., range, speed, or power. This will progressively improve in performance.

Are total-electric vehicles pollution-free? There is no exhaust issuing from the powered vehicle. However, to generate the electricity to charge the batteries to make the car's motor power up the vehicle, nonrenewable energy is used at power plants to generate the needed electricity. The power plant produces stack emissions that pollute the atmosphere. Alternately, we can use the burning of fossil fluid fuels (gasoline, diesel) to run a combustion engine in a vehicle that exhibits high performance. For now, these vehicles, although their engines are tuned for optimum performance and minimum pollution, still produce carbon monoxide, nitrogen oxides, particulates, and volatile organic compounds that are emitted at the tailpipes. In the future, we should look forward to powering vehicles or recharging batteries with renewable energy sources that, also, do not pollute.

Aside from pollution, our present-day generation of electricity is nonrenewable. Nonrenewable energy resources when used up are essentially gone forever. Those energy resources were developed naturally over millions and millions of years and cannot be renewed in our lifetime or the lifetime of millions of generations in the future. Although we keep finding new, nonrenewable sources, our booming population of humans will deplete these sources at accelerating rates of use. We expect reasonable quantities and prices of fuel (as at the gas stations) without thinking about the realities of finite sources, just like we expect to live forever.

Fossil Fuels

Fossil fuels[2] are solar fuels developed millions and millions of years ago. They are called fossil fuels because they have been formed from previous life forms and contain the energy gathered prior to their burial.

Coal

The Sun has been beaming energy towards Earth for about 4.6 billion years (Earth's estimated age). About 2.5 billion years ago, plants, by photosynthesis, produced enough oxygen to support life. About 250 million years ago (*Pennsylvanian Period*), plants became so abundant on land that forests flourished and plant matter accumulated in abundance. Photosynthesis captured energy from the Sun and stored it in the atomic bonds holding the glucose (sugar) molecules together. Burial of these forests by overlying sedimentary sequences created a "burial chamber" to make coal. Buried with the coal are remnants of the bygone forests—plant fossils. Later, other forests were buried during the *Tertiary Period*.

During deep burial for long periods of time, heat and pressure drove off some of the volatile constituents of the plant matter and concentrated the carbon. The plant material went through a metamorphism from pure plant detritus, to *peat*, and then to various grades of coal: brown coal (*lignite*), to soft coal (*bituminous*), and in some cases to hard coal (*anthracite*), concentrating the energy more and more with each step[3] **(Figure 15-1)**.

FIGURE 15-1 Coal

Lignite deposits developed in the Tertiary strata. Bituminous deposits developed in Pennsylvanian (*Carboniferous* overseas) strata. We can burn these products today and recapture some of that stored energy. The higher the grade of coal, the less solid waste (ash) is produced and the cleaner the burn.

Near-surface coal deposits can be mined by surface mining (*strip mining*) methods. This involves stripping off the *overburden* of soil and overlying rock. Then, *drag-lines* (huge mechanical shovels) scrape the coal out of the ground. After mining, current regulations require the coal company to put back the overburden rock first, then the subsoil. The surface is graded, so it drains but is not prone to precipitation runoff erosion. Then, topsoil is placed and seeded for reuse of the land (usually for grazing). Prior to putting these regulations in place, coal companies merely abandoned the stripped land, which was an eyesore and had no possibilities for future use.

Deep coal deposits have to be exploited by *underground mining* methods. First, vertical (*shafts*) or horizontal (*adits*) entryways are drilled, blasted, and dug to the level of the coal seams. A series of mine tunnels is extended along the beds of coal. *Room-and-pillar mining* was used in the past, excavating rooms and leaving pillars of coal to support the roofs. To prevent collapse while miners excavated the coal from each room, roof bolts and timbers were also installed between the pillars. When the mining operation was completed, the voids created were abandoned. Unfortunately, in time, the pillars and timbers decayed and collapsed and the roof bolts became loose. Without any support of the mine roofs, they collapsed, creating voids above. These movements worked their way to the surface. Since the mining, houses, businesses, roads, and other infrastructure had been built at the surface. The resulting subsidence damaged or completely destroyed many of these structures. As the damage usually occurs long after the coal companies have ceased to exist (disincorporated 50 to 100 years ago), no one is available to take responsibility for damages.

To avoid this problem for current mining operations, a new method of underground mining was developed. *Long-wall mining* temporarily supports the roofs using hydraulic roof supports. Mechanized excavators tunnel to the end of seam, excavate the coal, and move it out on conveyor belts as the machines withdraw, removing the roof supports as they retreat. Subsidence occurs instantly and the damaged parties above are compensated quickly for their losses.

Coal as a fuel is used to produce heat to generate electricity and to process metallic ores. In the past it was used as the main direct source of heat for heating residences and cooking. Because of the relatively high air pollution caused by burning coal, residences in the United States are now using cleaner fuels (gas and fuel oil) for direct heating and cooking. Residences, however, also use electricity for heating and cooking, although the electricity is generated at power plants, many of which still use coal. Coal still has widespread use in developing countries, where coal deposits are abundant (e.g., China).

Petroleum

Another important fossil fuel source is petroleum, from which we get gasoline, diesel, airplane, and other fuel formulations to power our vehicles and to run other machines[4] **(Figure 15-2)**. As these are liquid fuels, they can be used effectively for transportation. *Petroleum*, literally oil (*oleum*) from rocks (*petros*), is derived from microscopic animals and plants that lived millions of years ago. Their remains (also containing energy from the Sun) were naturally processed over long periods of time, buried deeply beneath the

FIGURE 15-2 Pumping crude oil

Earth's surface, where chemical reactions converted their remains into a substance called *kerogen*. Kerogen was converted to *crude oil* (*crude*) and natural gas (*methane*). The crude migrated through the permeable rock strata to natural collection areas (*pools*), having been trapped in porous rock against adjacent impermeable rock strata. Migrating with the crude and gas is saline water (*brine*). Density of the three fluids (brine, crude, and natural gas) from greatest to least stratifies the pool with brine at the bottom, crude above the brine, and natural gas on top. Brine is wastewater that must be properly disposed of if pumped from the ground.

Crude oil is sent to refineries to fractionate the mixture into various fuels (e.g., gasoline, diesel, fuel oil) and nonfuel products (e.g., plastics, medicines, lubricants, food dyes). The natural gas (methane) is cleaned (sometimes extracting metals) and sent through pipelines to their point of use (e.g., electrical generating power plants, municipal buses, buildings for heating and cooking). Petroleum fuels are generally cleaner burning than coal, and natural gas is cleaner burning than liquid petroleum fuels. Natural gas had been flared (wasted it at the wellhead) because it had no market or network of pipelines to move it to market; now it is used extensively.

A relative newcomer to nonrenewable, natural gas reserves is methane from shale. *Hydro-fracturing* (commonly referred to as *fracking*) and *horizontal drilling* are techniques that can release marketable methane from shale from large subsurface areas from one wellhead location. Millions of fractures are produced in the rock by injecting water, chemicals, and sand into shale formations under high pressures. This practice is controversial because of the danger that methane and fracking fluids increase the flow of methane into nearby water supply aquifers. Also, it uses great quantities of fresh water and produces great quantities of recovered water that is contaminated with fracking chemicals. Some of these waste fluids have been injected into deep wells that have triggered earthquakes, and some waste fluids have been illegally dumped into sewers leading to freshwater streams.

NUCLEAR ENERGY

Nuclear energy[5] is also nonrenewable as it relies on a finite amount of *uranium* ore on Earth for processing nuclear fuel. Most uranium compounds on Earth contain the uranium isotope, U-238 (92 protons and 146 neutrons) with minor amounts of U-235

FIGURE 15-3 Nuclear power plant, Spain (notice, there are no cooling towers—plant is adjacent to the sea for recirculation of cooling waters)

(92 protons and 143 neutrons). Note that the addition of protons and neutrons equals the atomic weight of the uranium, indicated after the dash, i.e., U-235 and U-238. It is U-235, contained in the ore as a minor constituent, which is fissionable fuel, valuable in a nuclear chain reaction to generate electricity. Consequently, the ore must be concentrated, using centrifuges that separate the heavier isotope (U-238) from the lighter one (U-235). U-235 fuel pellets are used in a nuclear reactor to generate heat, boil water, generate steam, turn turbines, and make electrons flow through wire. This is similar to fossil fuel (coal-fired or gas-fired) power plants, except instead of burning fossil fuels for the heat, a nuclear reaction is used to produce the heat **(Figure 15-3)**.

Another big difference between fossil fuels and nuclear power plants is the different wastes produced by each fuel. Coal-*fired power plants* that generate electricity produce air pollutants and solids in the form of coal ash (*bottom ash* and *fly ash*). Air pollutants are cleaned (a process, called *scrubbing*) prior to emission from chimneys (*stacks*), including the light fly ash. Coal ash can be landfilled or mixed with cement for concrete production. Some of the ash has been used to backfill abandoned underground mines, in voids produced by the mining of the coal that generated ash. *Gas-fired power plants* that generate electricity produce less air pollution than its coal-fired counterparts, and hardly any solid waste. The fuel, however, is more expensive at present, but increases in gas production may make it economically viable to convert power plant coal furnaces into gas furnaces and reduce waste management costs.

Commercial *nuclear plants* that generate electricity in the United States, numbering 65 in 31 states have from one to three units per site (104 units total) and account for about 20 percent of the electricity generated. The waste that nuclear plants produce is mostly radioactive. The waste includes spent fuel pellets, fuel rods that control the reaction, and water that the rods are bathed in to control the chain reaction. The *half-lives* of the radioactive material are so great (thousands of years) that it must be isolated from human exposure. As there is no central repository for commercial nuclear waste to date, the waste products must be stored at each of the plants, protected from radiation exposures and leaks. Reactor safety precautions and radioactive waste storage are costly, limiting the benefits of power generation efficiency using nuclear fuels.

The generation of electricity by nuclear plants was once considered (ex-President Eisenhower's Atoms for Peace Program in the 1950s) a cheap (or even free) source of energy. In reality, nuclear power has not competed well economically with fossil fuels.

Hydrogen Fuel Cells

Although hydrogen fuel cells are an alternate energy source, for now they should not be characterized as a renewable source. This is because the hydrogen needed for the fuel cells is currently produced from fossil fuels (nonrenewable). In the future, if we can extract hydrogen from water using renewable energy sources, hydrogen fuel cells can be reclassified as renewable.

Fuel cell vehicles have potential if hydrogen gas can be produced economically. Hydrogen is passed through a PEM (*proton exchange membrane*) fuel cell.[6] This separates each hydrogen atom into a hydrogen ion and an electron. Electrons are diverted through a wire to create an electrical current (a flow of electrons). After energy use, the hydrogen ion, the electron, and an oxygen atom join to form the only waste product, pure water. Hydrogen gas can be made by processing natural gas (methane) or other hydrocarbons by high temperature steam reforming. To be practical as an auto fuel, hydrogen dispensing stations must be readily available. The electrical energy produced should be greater than the energy needed to produce the hydrogen gas.

References

1. Barnett, D. 2000. *Electric Power Generation: A Nontechnical Guide.* Tulsa, OK: PennWell.
2. Grant, L. 2005. *The Collapsing Bubble: Growth and Fossil Energy.* Santa Ana, CA: Seven Locks Press.
3. Thomas, L. P. 2013. *Coal Geology.* Hoboken, NJ: Wiley-Blackwell.
4. Montclaire, K. L. 2008. *Petroleum Science Research Progress.* New York: Nova Science Publishers.
5. Miller, D. A., ed. 2010. *Nuclear Energy.* Detroit: Greenhaven Press.
6. Larminie, J. 2000. *Fuel Cell Systems Explained.* New York: Wiley.

16 Renewable (Sustainable) Energy

Our enormous population and its infrastructure demand more energy. We cannot recycle energy. We can only, during its use, degrade it into less useful forms (*entropy*). Society refuses to significantly reduce our thirst for energy by returning to the lifestyle of historic and prehistoric times. Therefore, we must put in place sustainable, *renewable energy* reserves quickly, before nonrenewable sources dwindle to naught.

Renewable energy sources will be available for the duration of human existence on Earth, and maybe beyond. This source is sustainable because no matter how much we use, renewable energy supplies will outlast us. This includes solar, hydroelectric, wind, biomass, biofuels, tides, and other forms that we have not yet thought to harness. Before we completely abandon fossil fuels, there has to be additional technological development of sustainable sources to generate enough energy at affordable prices. As an added benefit, those sustainable sources mentioned produce less pollution and danger than fossil or nuclear fuel sources.

To provide enough energy for our accelerating population growth and the technological infrastructure, which we have come to need and enjoy, we look to the Sun for additional energy. There are many ways to tap more energy, some of which we are slowly realizing.

Directly or Indirectly from the Sun

Nature has harnessed the Sun's energy in many ways.[1] Plants, with the help of chlorophyll, use this energy to perform photosynthesis from the visible light that the Sun provides. Our atmosphere, with the help of *greenhouse gases*, uses the Sun's *radiant* heat energy to warm the planet to an average global temperature of 57.2°F (+14°C), much higher (60 Fahrenheit degrees higher) than the 1.4°F (−17°C) if greenhouse gases were not naturally present (i.e., without an atmosphere), as calculated by Joseph Fourier[2] in 1824. Gravitational attraction of the Sun on the Earth's oceans has helped (with gravitational attraction of the Moon) to move ocean water onto and off of the continents daily or twice a day to support intertidal and brackish-water habitats on continental shelves, in estuaries, fjords, embayments, and beaches. Ultraviolet rays from the Sun in moderation add vitamin-D to our bodies through reactions in our skin. Solar heat evaporates water from the sea, lakes, and soil to circulate moisture, purifying and distributing water for the hydrologic cycle. Uneven heating of the atmosphere, together with the axial rotation and orbiting of the Earth, cause weather systems to move, cause winds to blow, and seasons to alter our environments. Last, but not least, the light of the Sun during the day and reflected light at night allow us to experience the sights of our surroundings and help to set our internal biological clocks to regulate our biorhythms.

Mimicking nature, we can tap the Sun's energy for our increasing energy needs. Most of the solar methods optimally function in areas of sunny climates, but all areas can benefit, at least to supplement other sources. No pollution or waste products are produced. Energy can be used locally without the need for long-distance delivery systems.

Solar Cells

Photoelectric cells[3] take the incoming energy of the Sun and directly convert it into electrical energy **(Figure 16-1)**. These cells can be connected into an array in panels to multiply the amount of electricity they can generate. Panels can be mounted on roofs or on the ground. In rural areas, though, the array of panels uses land that could have other uses.

Active Heat Collectors

Parabolic reflectors can collect and concentrate incoming solar radiation. You may have concentrated a beam of the Sun's rays by using a magnifying glass (to start a campfire). Similar to the actions of power plants, the concentrated light can heat fluids that can turn turbines and generate electricity.

FIGURE 16-1 Solar panels of photovoltaic cells

Passive Heat Collectors

Passive heat collectors work by exposing glassed-in indoor areas to the Sun's rays during the day (like a greenhouse) and heating up ceramic floor tiles. The windows should face a direction that has maximum solar exposure (southern in the northern hemisphere and northern in the southern hemisphere. At night, the heat reradiates up from the warmed floor to heat the air in the room. Insulated drapes that cover the windows at night help to keep the warmed air from escaping. To optimize this method, such rooms should have maximum solar exposure with a row of deciduous trees outside—shading the room in summer with trees fully leaved and allowing sunlight to penetrate in winter after leaves have fallen.

HYDROELECTRIC

Because the Sun has a major influence in the hydrologic cycle, it can be credited for producing hydroelectric power. The Sun evaporates the water and lifts it to high elevations, providing the potential energy to initiate hydropower **(Figure 16-2)**.

Potential energy of water behind a dam can flow through its *penstocks* (*sluice gates*) and turn water wheel turbines, thus generating electricity without burning fuel or steam production.[4] Engineers should be wary of placing dams at the narrowest width of a stream valley, as these areas (requiring lesser amounts of materials to cover the span) may be narrow due to tectonic structures in the rock. Faulting, jointing, and intrusive rocks may be tectonic structures that are located at these stream reaches. These dam sites may require grouting prior to construction to provide good foundation conditions and impermeability for the dam to function safely.

Although no pollution is released during hydroelectric power generation, the construction of the dam changes natural environments from *terrestrial* (land) and *fluvial* (stream) to *lacustrine* (lake) habitats, displacing aquatic and terrestrial wildlife in the former undammed stream. Dam safety may also be an issue, with habitation downstream in danger during flooding and dam breaks. The lakes behind the dams require periodic *dredging* from the collection of sediment that does not enter the penstocks. If dams are in remote areas, high-tension electrical transmission lines must be constructed to transmit electricity to populated areas.

FIGURE 16-2 Hydroelectric dam, Tolliatty City, Russia

Wind

Because the Sun has a major influence in producing wind by unequal heating of the atmosphere, it can be credited for helping (with contributions from the Earth's axial rotation) to produce wind power.[5] Wind directions are influenced by the resulting rebalancing of high- and low-pressure systems, which redistribute masses of air. Some geographic areas are windier than others. Lines of giant wind turbines are erected to catch the breezes and turn the blades of these units. No pollution is generated. However, the blades sometimes interfere with migrating flocks of birds. These huge structures, furthermore, disturb the view of picturesque landscape and prevent their use in some touristic areas.

Biomass

Used as a fuel, *biomass* needs to be burned to extract energy as heat.[6] Biomass consists of waste organic matter, such as plant litter, animal feces, food wastes, and trash. It is usually not processed, so nonburnable materials mixed in create lots of material that needs to be disposed of (in a landfill). Moisture also inhibits complete combustion. Unlike coal, oil, or natural gas, volatile materials have not been driven off; so the amount of heat per weight is less than those of fossil fuels. Advantages include low (or no) cost and volume reduction of trash targeted for landfills. Biomass is sustainable, as humans have no trouble in producing organic wastes.

A biomass product that has been used by primitive and poor societies for heating and cooking is wood. Wood is only semisustainable because some trees take many years to grow back. Western Europe's timberlands were depleted centuries ago, as evidenced by the relative scarcity of wooden houses and power/telephone poles in some countries. In Korea, country dwellers have used charcoal to heat homes by warming the floors using charcoal below. Unfortunately, burning charcoal can produce carbon monoxide that could seep up between floorboards, asphyxiating sleepers in rooms with poor fresh-air ventilation.

Bricks of animal manure have been used on farms in the French Alps, first as roof insulation and storage in the summer—then as fuel for heating barns.

Biofuels

As with other fuels, biofuel needs to be burned to extract energy as heat.[6] Biofuels are processed from plants like corn, sugar cane, and other annual crops (sustainable from year to year). This fuel, therefore, is another example of energy from the Sun through plant photosynthesis. The fuel is *ethanol (ethyl alcohol)* and can be added to gasoline up to 20%, or with engine modification as a straight fuel. Other examples of biofuels (used after engine modification) are algae "soup" or used cooking oil.

Ocean Tides, Waves, and Currents

The ocean is another of Earth's powerhouses.[7] The gravitational pulls of the Moon and Sun move seawater onto and off the continents daily or twice a day on continental shelves, in estuaries, fjords, embayments, and beaches. When tides pull ocean water up gentle slopes into stream valleys, a tidal bore can develop that is strong enough to turn turbines and generate electricity. An added feature of this system is realized when the tide goes out; the turbines can turn again and generate more electricity. Two electrical power plants have been built in France and the United States using energy generated from tides.

Ocean wind waves are also a manifestation of solar energy (from wind). The motion of floating pipe segments joined by wires in wave action can also generate electricity. Ocean currents powered by differences in water temperature and salinity can also be harnessed to capture this energy.

Geothermal

One of the few energy resources not linked to the Sun is from the heat generated from natural nuclear reactions within the Earth's crust.[8]

Certain hot spot locations on Earth near volcanic activity have naturally heated groundwater. Examples include a spring of this water, called a *geyser* (e.g., Old Faithful in Yellowstone Park) or hot mud pools (e.g., Solfatara, Italy). Iceland, with numerous volcanoes, that sits on the Mid-Atlantic Ridge (boundary between the North American and Eurasian tectonic plates) has enough naturally heated water to heat all its homes without using other forms of energy.

In New Zealand, the central area of its North Island has the most exploitable geothermal sources. The eastern edge of the island sits on a plate boundary (between the Australian and the Pacific tectonic plates). Its geothermal heat can generate between 10 and 15 percent of the country's electricity.

Geothermal energy in non–hot spot areas can also contribute to our energy needs. The normal geothermal gradient (warmer at progressively greater depths) away from plate boundaries is about 25 Celsius degrees for every kilometer of depth (45 Fahrenheit degrees for every 0.6 miles of depth). At shallow depths, groundwater at a temperature of 55°F (12.8°C) can be pumped and circulated into homes at the surface to lower room temperatures (with dehumidification) to comfortable levels in summer and, with supplemental heating, raise room temperatures to comfortable levels in winter. This system must use electrical energy to pump and circulate the waters. It does not pollute and is renewable.

References

1. Richards, J. *Solar Energy.* Tarrytown, NY: Marshall Cavendish Benchmark.
2. Cowie, J. 2007. *Climate Change: Biological and Human Aspects.* New York: Cambridge University Press.
3. Green, M. A. 2000. *Power to the People: Sunlight to Electricity Using Solar Cells.* Sydney: University of New South Wales Press.
4. Draper, A. S. 2003. *Hydropower of the Future: New Ways of Turning Water into Energy.* New York: Rosen Publication Group.
5. Rivkin, D. A., M. Randall, and L. Silk. 2014. *Wind Power Generation and Distribution.* Burlington, MA: Jones & Bartlett Learning.
6. Smil, V. 2013. *Harvesting the Biosphere: What We Have Taken from Nature.* Cambridge, MA: MIT Press.
7. Gerdes, L. I., ed. 2011. *Wave and Tidal Power.* Detroit: Greenhaven Press.
8. DiPippo, R. 2005. *Geothermal Power Plants Principles: Applications and Case Studies.* New York: Elsevier.

UNIT 5 Environmental Problems

Waste and Pollution

We use and discard materials.[1] If we do not need those materials, they take up space. A space problem is minor. What is more significant is potential environmental impact when the waste has *hazardous chemicals*, *particulates*, or *pathogens* (organisms that can cause disease). These chemical and particulate waste products can pollute our air, water, and soil.[2] Pathogens (e.g., some bacteria, viruses, fungi, protozoans) can cause disease. Waste can also attract *vectors* that can spread disease.

Sources of Waste

Manufacturing

We manufacture articles and chemicals and, thereby, generate wastes, some of which have hazardous characteristics. Potentially hazardous products must be tested prior to release to consumers. In the manufacturing process some of these products generate wastes that might include effluents, spills, stack emissions, fugitive emissions (like dust from construction), *stack scrubber wastes* (from trapping stack emissions), scraps, sludge, spent containers, and worn-out parts. Potential future wastes from manufacturing are the products that have been used and are now obsolete and the packaging in which the original products were shipped in.

Agricultural

Modern farming uses chemical fertilizers and pesticides. Excess chemicals can run off the land and pollute streams and lakes. Feedlots where cattle are fattened before slaughter produce a concentrated supply of manure and excess cattle food that is washed into streams and lakes. Worn-out parts, farm equipment, and vehicles can also add to the amount of waste generated by agriculture.

Resource Extraction (mining, drilling, and processing)

Resource extraction activities pose safety risks to workers from flammable-gas mine explosions, mishandling of blasting explosives, cave-ins, and inhalation of dust and fumes. Waste products from mining and drilling also present health problems to both miners and the public.

Mining probably produces the most waste by volume and weight than any other source (**Figure 17-1**). This is because what is extracted from the Earth contains a high percentage of unusable rock material. This so-called *spoil* ends up in piles at the surface or so-called *gob* underground. Spoil from coal mining can ignite spontaneously and cause fire and air pollution hazards. *Leachate* from effluent generated in spoil piles is also a problem waste. Leachate generated in abandoned underground coal mines can produce acid mine drainage. Mining operations can also endanger the public when toxic and explosive gases migrate to areas adjacent to mining operations. Surface mining can release fugitive dust emissions into the air in adjacent areas.

Drilling operations generate wastes. In drilling for oil and gas or water, cuttings, used drilling muds, and brine (salt water) can be generated. Containment and treatment of these muds and brines are crucial; otherwise they can pollute environmental media.

Fluids injected into subsurface rock formations may also create environmental problems. *Deep-well injection* can pollute subsurface aquifers if the injection wells are not

FIGURE 17-1 Mining wastes

properly sealed. Deep-well injection has been blamed for triggering low-Magnitude earthquakes when the injection wells are located near weaknesses in the Earth's crust (fault zones).

Oil and gas operations, using a process called *hydraulic fracturing* (or simply *fracking*), involves the injection under hydraulic pressure of water, sand, and chemicals. The injection process fractures shale formations that contain natural gas (methane) and crude oil, releasing them for extraction and exploitation. In this process, wastes are generated from injected fluids that return to the surface. To avoid release into environmental media, boreholes must be sealed when they penetrate aquifers. Improper sealing of the annuli of boreholes can cause some of the circulating fluids to laterally invade groundwater aquifers. Disposal of these return liquids by injection into deep wells away from the fracking operations must also be mindful of the need to use disposal wells that are properly sealed. Wastewaters from fracking can also be released into the environment through spills and leakage from improper liquid storage, resulting in the release of treated or untreated wastewaters into surface waters. Regulated release of wastewater is allowed if permits are obtained under the USEPA *National Pollutant Discharge Elimination System (NPDES)*.

Processing of mined-out materials and fluids can also be environmentally problematic. Smelting of ores can create gaseous emissions and waste solids. Petroleum refineries and chemical plants using petroleum products can release contaminants to environmental media. Heap-leaching of ores (in fluids such as cyanide solvents) in improperly lined or located containment ponds can also cause environmental problems.

Like agriculture, mining and drilling operations generate waste in the form of abandoned worn-out equipment and parts, vehicles, and ore handling facilities.

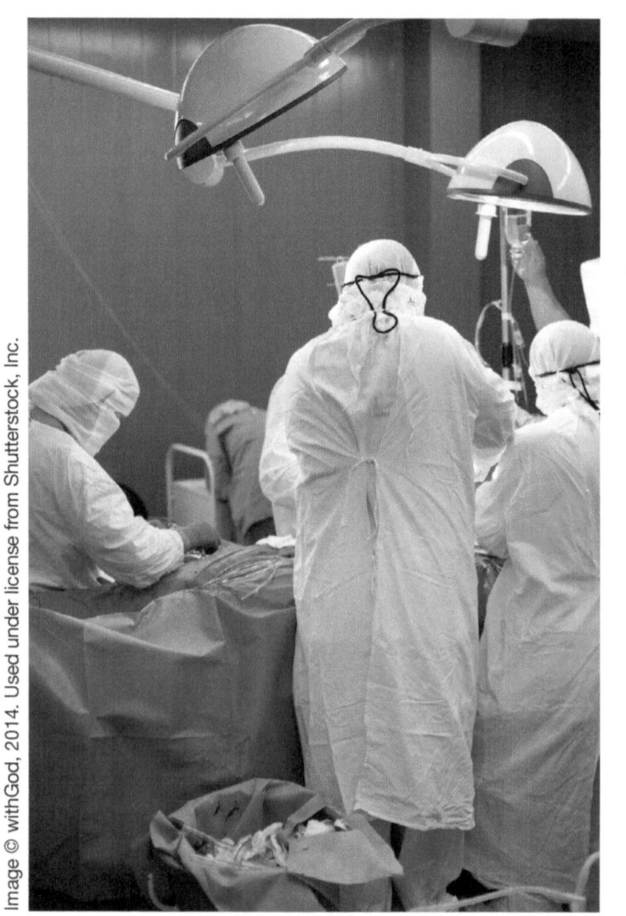

FIGURE 17-2 Medical waste generated during surgery

Medical

Medical materials and wastes can include used bandages, pharmaceuticals, radioisotopes, packaging, chemicals, laboratory cultures, food, blood, tissues, specimens, disposable instruments, so-called *sharps* (e.g., syringes, needles), and wipes (**Figure 17-2**). These are potential biohazards because such discards can transmit diseases.

Transportation

Public or private transportation vehicles use fuels that are flammable, explosive, and toxic. The handling and transportation can result in spills, exhausts, lubrication drips, and fuel leaks along transportation routes. Accidents of vehicles carrying fuel products and wastes can increase the risk of these wastes being released into the environment.

Spent batteries, especially from vehicles, are waste products traditionally containing both lead (electrodes) and corrosive battery acids. Electric car or hybrid electric car batteries can contain other metals like nickel, lithium, cobalt, and other metals.

Power Generation

Electrical power generating stations use fuels and release wastes. A fossil fuel plant emits gas and particulates from the *combustion process* that gets vented by stacks. *Scrubbing* (precipitating waste prior to exiting the stack) produces solid waste (*scale*) from this operation. The combustion process produces both heavy *bottom ash* and lighter *fly ash*, which usually contain concentrations of heavy metals. Water used to cool the combustion process is sent while hot into streams, bays, and lakes.

If the plant is nuclear, radioactive materials and wastes might include fuel rods with radioactive fuel pellets, water from pools in the reactor core, and other "hot" reactor components. Low-rad wastes can also include wipes, lab coats, and gloves.

Commercial

The workplace generates and stores a variety of hazardous materials and wastes depending on the products or services provided. These may include cleaning fluids and other solvents, packaging, office litter (including paper), food and food wastes, residual chemicals in containers (**Figure 17-3**), metal plating wastes, paints, demolition and construction wastes, fugitive emissions from earthmoving, and scrap metals, among others. Some of these wastes might be suitable for disposal in a municipal solid waste landfill, while others might be better placed in a hazardous waste landfill or require incineration or processing prior to final disposal.

Residential

At home we store products or generate wastes that may be hazardous, such as refuse, food waste, water from toilets and plumbing fixtures (sewage), household chemicals, garden and lawn chemicals, insecticides, discarded furniture, parts, and equipment, lubricants, pharmaceuticals, toiletries, paints, and tobacco smoke.

FIGURE 17-3 Deteriorated chemical waste drums

FIGURE 17-4 Recycling (glass) bin, Sicily, Italy (photographed by Alan Jacobs)

FIGURE 17-5 Recycling collection, northern Italy. The bin can be hydraulically lifted and emptied (through its trap door) into a truck by one operator (photographed by Michael Bolan, M.S. Thesis, Youngstown State University)

Each person in a household produces an average five pounds of solid trash and 70 gallons of sewage per day. This does not include all the solid and liquid wastes generated in producing products, services, and energy for the consumer (a personal waste footprint).

The more a household recycles **(Figures 17-4 and 17-5)** or reuses articles that would have been put in the trash, the lower the generation of solid trash (destined to be landfilled). Water use economy would also lower the amount of wastewater generated.

References

1. Vaughn, J. *Waste Management: A Reference Handbook.* Santa Barbara, CA: ABC/SLIO.
2. El Nemr, A., ed. *Impact, Monitoring, and Management of Environmental Pollution.* New York: Nova Science.

Chemicals of Concern

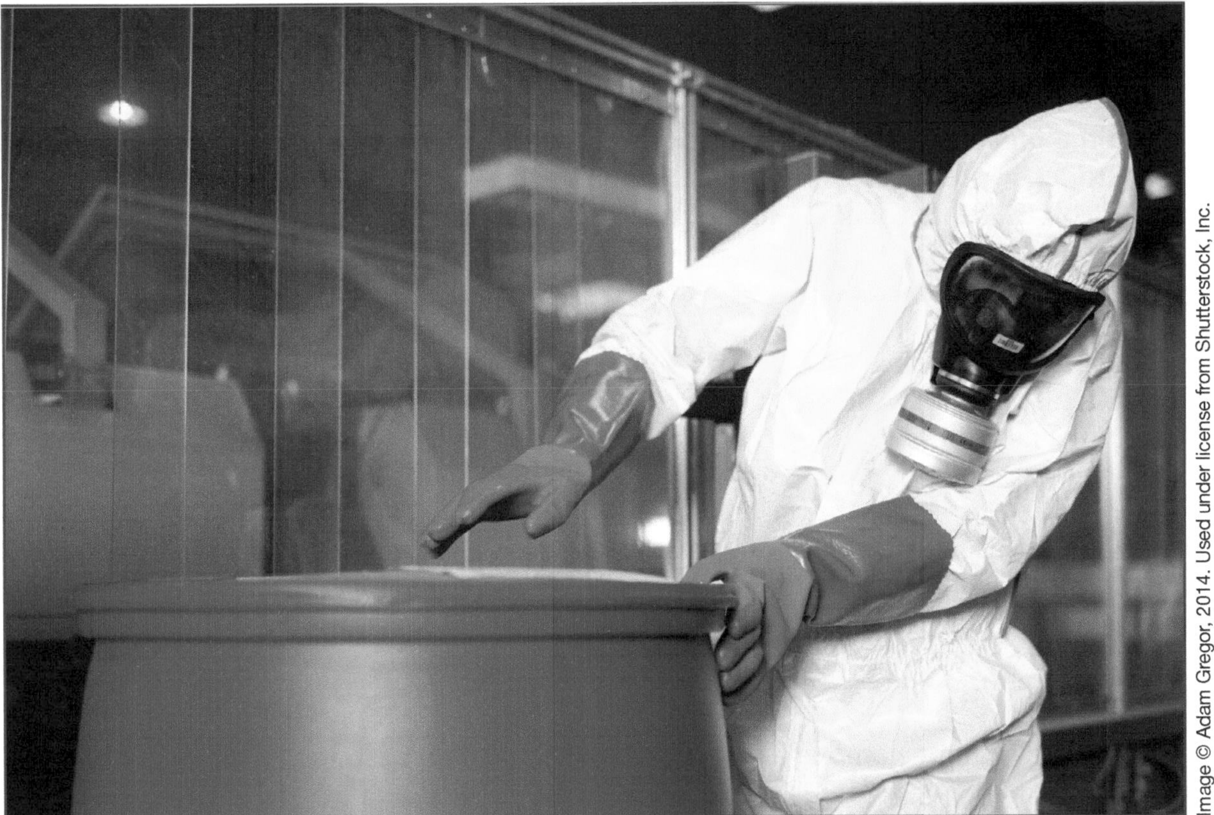

The Universe, including our Earth, is composed of *chemicals*. We are chemicals. Although chemicals differ in their toxicity (toxins are poisons), depending on the dose, all chemicals can be toxic.[1] Even essential minerals that are vital to our health in low doses can be harmful or lethal at higher doses. Even pure water can be toxic if ingested in such great quantities that it dilutes our electrolytes.

To be hazardous, chemicals must be present in sufficient volumes, have virulent toxic properties, be delivered to receptors through environmental pathways, keep their concentrations and toxic properties during transport along the pathways, and target receptors that are exposed to the chemical. Therein lies the solution to chemical hazards—namely, change this chain of events so that the chemical has reduced volume, toxicity, and mobility. Unfortunately, that may not be as simple as it sounds.

The concentration of the chemical may increase or decrease at its source, along its pathway, or after entry into the body. Source concentrations could change by dilution or evaporation. Pathway concentration could change from dispersion, dilution, mixing, or chemical reaction. Point of exposure concentrations in the environmental medium could be different from the concentration in the receptor tissues because of difference in solubility (receptor swimming in water vs. lipids in his/her fatty tissue). Contaminants may enter the body through different pathways and target certain organs in the body. Chemicals may be eliminated, metabolized, or stored in the body. Effective doses depend on body size, target organs, and rates of absorption.

Chemicals that are common contaminants could be inorganic (mostly metals) or organic (in general, containing carbon). They may cause cancer (*carcinogenic*), or not. They may produce birth defects (*teratogenic*) or damage genes (*mutagenic*). Their exposures and effects may be in large doses over short periods of time (*acute*) or in small doses over long periods of time (*chronic*). Toxicity of a chemical varies widely; the USEPA provides toxicity data on its Integrated Risk Information System(IRIS) website.

The production and use of chemicals have improved the quality of life. They are essential to the functioning of modern society. The proliferation of man-made chemicals and lack of proper waste disposal, however, have produced health hazards that also degrade the quality of life. What chemical should we be aware of and how do they affect our health?

Toxic Metals

The following metals pose health concerns[2]: *cadmium (Cd), chromium (Cr), nickel (Ni),* mercury (Hg), beryllium (Be), lead (Pb), and arsenic (As).

Toxic metals cannot be destroyed, but can only be isolated from the environment. Remediation methods that can treat organic compounds, including chemical treatment, bioremediation, and incineration, cannot lessen the toxicity of metals. They may even combine with organic compounds to make them even more toxic or their compounds can break down in acidified waters (from *acid rain*) to make them more mobile and soluble—and thus easier to be absorbed or ingested by receptors.

Cadmium (Cd)

Cadmium is important commercially. It is used in nickel–cadmium batteries, pigments, coatings, and electroplating. It is found in cigarette smoke and some foods. Cadmium bioaccumulates in seafood. Health effects include osteoporosis in women, height loss in men, kidney damage, elevated blood pressure, cardiovascular diseases, and Itai-Itai disease characterized by bone softening, severe pain, and kidney damage.

Chromium (Cr)

Chromium compounds were used in corrosion inhibitors, dyes and paints, and the tanning of leather. Chromium VI (valence of 6) is a carcinogen. It was mentioned as the cause of health problems in Hinkley, California (subject of the film *Erin Brockovich*). Chromium VI can cause kidney and liver damage, skin ulcers, and respiratory problems.

Nickel (Ni)

Nickel is used in nickel–cadmium batteries and, when added to other metals (creating alloys), increases their corrosion and heat resistance. Health effects from exposure include contact dermatitis and, when ingested, heart and liver disease. It is also a carcinogen.

Mercury (Hg)

Elemental mercury is liquid at room temperature. It has been used in thermometers, medicine, amalgam dental fillings, cosmetics, and light bulbs.

Organic mercuric compounds are more toxic and water soluble than elemental mercury, and thus can do more damage. A tragic case of note started with the discharge of inorganic mercury wastes from a plastics factory in Minamata Bay, Japan. The mercury was complexed by bacteria into highly toxic methyl mercury, which bioaccumulated through the aquatic food chain; humans and their pet cats then ingested the seafood. Although early warning signs were misinterpreted when the cats acted deranged, the problem was not linked to eating seafood until after human deaths, neurological diseases, and birth defects were widespread in this area.

Beryllium (Be)

Beryllium is used in industry because it is lighter than aluminum and stronger than steel. Exposure by metal processing workers via inhalation can lead to cancer.

Lead (Pb)

Lead has been used in gasoline, paints, pottery, auto batteries, and solder. In the United States lead has been removed from currently manufactured products. Older houses, however, can still have lead-painted surfaces that chip and form paint dust. Exposure can lead to adverse health effects to the central nervous system. It is a common natural component of soils, whose background concentrations depend on the parent rock from which the soil was derived.

Arsenic (As)

Arsenic is also a common natural component of soils. The natural background concentration depends on the bedrock parent material from which the soil was formed. Rocks containing the minerals realgar and orpiment (arsenic sulfide) can weather and incorporate arsenic into the soil. Arsenic is used as a wood preservative, in medicine, and in metal alloys. Exposure to arsenic can cause liver, skin, bladder, and kidney cancer and other diseases.

Aluminum (Al)

Aluminum is used in construction and manufactured products because of its light weight. It is also used in food and beverage containers, wraps, and cookware, resulting in its potential to be ingested. It is ingested from medicines that contain aluminum (buffered aspirin). Dermal contact with cosmetics and antiperspirants that contain aluminum can also subject our bodies to exposure to this metal. Small amounts of ingested aluminum (which can be excreted) may not result in adverse health effects. There is a possible linkage between high aluminum retention in the body and Alzheimer's disease.

Potentially Harmful Metals We Cannot Live Without

Some metals, including copper (Cu), zinc (Zn), and iron (Fe)), are both essential as micronutrients (in small amounts) and toxic (in larger amounts). This fact underscores the admonition of Paracelsus, known as the "Father of Toxicology," who stated that all substances can be poisons, depending on the dosage. Metals cannot be manufactured in the body like many organic compounds, so they must be ingested in proper amounts.[3] If insufficient amounts are obtained in foods, then mineral supplements must be used. However, if copious amounts enter the body, adverse health effects and even death may result.

Uses of *copper* include electrical wires and plumbing, alloying with other metals (bronze = copper + tin), and inhibitors of mildew and insects. Copper is needed in the diet for proper growth, utilization of *iron*, enzymatic reactions, and other health benefits. During the manufacture of copper products, copper-containing wastes are released into environmental media, which form pathways to receptors. Receptors who inhale, ingest, or come into dermal contact with copper can develop adverse respiratory effects and, at greater intakes, liver and kidney damage.

Uses of *zinc* include coatings and rust inhibitors, batteries, and alloys (e.g., brass = zinc + copper). Zinc is needed in the diet to maintain healthy immune systems and for child growth and development. Yet, higher dosages can cause gastrointestinal problems, anemia, and damage to the pancreas. An occupational disease called *metal fume fever* (resulting in flu-like symptoms) is caused by the inhalation of high concentrations of zinc (and some other metals).

Iron is the most important metal in construction materials and is used in the manufacture of heavy machinery, appliances, and vehicles. It is also needed in the diet for cell growth and the transportation of oxygen through the bloodstream. Excessive amounts of iron ingested, however, can be poisonous (acute iron intoxication).

Toxic Organic Compounds

Toxic organic compounds of concern[1] include *pesticides* (e.g., herbicides, *insecticides*, fungicides, rodenticides, nematocides), dioxins, polychlorinated biphenyls, solvents, and materials used in plastics manufacturing. Most of these compounds have been manufactured in industry (not found in Nature). New chemicals are of concern because long-term health effects are not yet known, and our bodies have not developed immunities (through natural selection) to these chemicals over the generations.

Pesticides include any substance(s) that prevents, destroys, or repels pests (insects, rodents, weeds, roundworms, and other organisms). Pesticides in the past included toxic metals or metallic compounds, but such pest killers have been phased out due to their toxic effects on humans.

Insecticides

Currently, insecticides[4] fall into the following groups of organic compounds based on their chemical structure: *organophosphates*, *organocarbamates* (*carbamates*), *organochlorides* (*chlorinated hydrocarbons*), and *pyrethroids*. They differ relative to persistence in the environment, range of targets (*broad or narrow spectrum*) that they can kill, and their short-term and long-term adverse health effects (**Figure 18-1**).

Organophosphates

Organophosphates include broad-spectrum insecticides, and do not persist in the environment. Yet, they frequently cause fatal poisonings, both acutely and in the long term. The insecticide works on humans in a similar way it destroys insects, by impairment of the transfer of nerve impulses. In the long term, it can cause numbness and weakness.

FIGURE 18-1 Spraying pesticides

Examples familiar to farmers include diazinon, malathion, and parathion, which can be administered to crops and soils in the form of granules and fumigants.

Carbamates

Carbamates, like organophosphates, also include broad-spectrum insecticides that dissipate quickly in the environment. They are used in household gardens to control insects and on dogs and cats to control fleas and ticks. Examples include *carbyl* (Sevin®), *aldicarb*, *fenoxycarb*, *propoxur*, and *metam sodium*.

In the manufacture of carbamates, a serious accident occurred in Bhopal, India, in 1984. An intermediate product in its manufacture, methyl isocyanate (MIC), was accidently released from a Union Carbide chemical plant in Bhopal and killed 3,800 people.

Metam sodium contaminated the upper Sacramento River in 1991 after a derailment of a Southern Pacific freight train. A tank car containing metam sodium plunged off a trestle and leaked 19,000 gallons into the channel. The chemical killed an estimated 200,000 fish and sickened 300 residents, who suffered headaches, chest pains, rashes, and nausea.

Organochlorides

Organochlorides are hydrocarbons containing chlorine atoms in addition to hydrogen and carbon. Unlike organophosphates and carbamates, organochloride pesticides do not break down easily in the environment and easily bioaccumulate. Most have been banned in the United States, including DDT (since 1972), chlordane (since 1988), methoxychlor (since 2003), and Mirex (since 1976).

DDT (dichlorodiphenyltrichloroethane) was a useful pesticide used to kill mosquitoes and control the spread of malaria. Although its use is banned in the United States and other developed nations, it is still used by many other countries in their food products and imported into the United States. DDT's main problem is persistence in the environment. If you spray a wall with DDT, this pesticide can still kill insects that land on the wall one year later. The main environmental concern is the thinning of the shells of birds' eggs. Human health problems from DDT include pancreatic cancer and non-Hodgkin's lymphoma, breast cancer, and reproductive and neurological disorders.

Chlordane has been banned in the United States since 1988. Between 1983 and 1988 it was allowed for termite control only. It was found to cause testicular cancer and other adverse health effects.

Methoxychlor was used as an insecticide on livestock, crops, and pets. It was banned since 2003 based on its acute toxicity, bioaccumulation, and potential for endocrine disruption.

Mirex is also banned in the United States because it is persistent in the environment and bioaccumulates in the body. It causes cancer in laboratory animals and, because of that, is considered to be carcinogenic in humans. Mirex is the contaminant of concern at a Superfund site south of Youngstown, Ohio, where this pesticide was manufactured.

Lindane, unlike the other organochloride pesticides, is still in use to kill lice and mites that attack the skin (and scalp). Side effects and allergic reactions may occur.

Pyrethroids

Pyrethroids are derived from a variety of chrysanthemum flowers. This plant species' natural insecticide has been harvested by humans. This insecticide paralyzes and kills flying insects by interfering with their nervous system. It is one of the safer insecticides, but it can adversely affect human nervous systems. Care must be taken when applied to crops or garden plants. Tingling and numbness indicate overexposure.

Herbicides and Defoliants

When the pest is a plant (e.g., weeds, poisonous plants), herbicides or defoliants are used.[5] Some are preemergent, which will prevent the seed from germinating, whereas others are postemergent, which kills the plant after germination. *Grassy weeds* (with blade-like leaves, such as crabgrass and annual bluegrass) are best controlled with preemergent herbicides. *Broadleaf weeds* like dandelion, poison ivy, and white clover can be controlled with postemergent herbicides.

Herbicides include chemicals such as *atrazine* and *paraquat*. Atrazine (applied only commercially by licensed users) kills broadleaf weeds and is applied pre- or postemergence. It has been linked to reproductive disorders, and may cause cancer in humans. Paraquat (applied only commercially by licensed users) kills broadleaf and grassy weeds. It can cause skin and lung damage.

An organic non-toxic herbicide that must be used as a preemergent herbicide (advertisements claim it is edible) is called *corn gluten meal (CGM)*. One could also try vinegar (acetic acid), which is effective on some weeds.

Defoliants are used to strip the plant of its leaves, rather than to destroy or inhibit plant growth. They are used in cotton harvesting. The defoliant *Agent Orange (2,4-D and 2,4,5-T)* was used in the Vietnam War. The main chemical used as the defoliant is still used today in commercial herbicides. But as used in the Vietnam War to defoliate trees that were harboring enemy snipers, the manufactured product contained dioxin impurities. Adverse health effects noted in soldiers returning to the United States (soft tissue sarcoma, non-Hodgkin's lymphoma, Hodgkin's disease, and chronic lymphocytic leukemia) were probably from the dioxins.

Nonpesticide Organic Contaminants of Concern

Aside from pesticides, organic compounds of concern to environmental health include dioxins, PCBs, solvents, and chemicals used in the manufacture of plastics (styrene and vinyl chloride).[1]

Dioxins

No one intentionally manufactures dioxins (they have no commercial use), but these chemicals are an unintentional byproduct of several industrial processes involving combustion at high temperatures, bleaching of wood pulp, and manufacturing of some herbicides. They are also produced when smoking cigarettes. Nature produces dioxins from forest fires and volcanic eruptions.

Of the more than 400 compounds in the dioxin group, only about 30 are considered poisonous. They are stable (persistent in the environment) and bioaccumulate in the food chain. When exposed, receptors can develop skin damage, growth of excessive body hair, liver damage, cancer, endocrine disruption, and reproductive and developmental effects.

PCBs

Polychlorinated biphenyls (PCBs) had two common uses prior to their being banned: transformer and capacitor fluids (insulators) and lubricants. In the environment, they can bioaccumulate in foods (fish and other animals) and, thus, can be ingested by humans. This contaminant has been identified at one-third of all the Superfund sites on the National Priorities List.

PCBs cause cancer in animals and probably in humans too. They impact the immune, reproductive, and developmental systems.

Organic Solvents

Industrial solvents at one time included alkanes, such as *benzene* and other flammable fluids. Benzene, the main component in gasoline, is a carcinogen. With the development of nonflammable chlorinated hydrocarbons, such as *trichloroethylene (TCE)* and perchlorethylene (PCE), industries using these solvents for cleaning were initially pleased about the new, presumably safer solvents. However, these chlorinated solvents presented two new serious environmental problems. First, TCE (used to clean circuit boards and machinery) and PCE (used as a dry cleaning fluid for clothes) **(Figure 18-2)** had their own adverse health effects. TCE and PCE have been linked to non-Hodgkin's lymphoma, leukemia, rectal cancer, bladder cancer, breast cancer, lung cancer, and adverse effects with fetal development. Furthermore, both PCE and TCE can degrade into vinyl chloride, which is a Class A carcinogen linked to liver and bile cancer in humans.

FIGURE 18-2 Garments that are dry cleaned

These solvents find their way into water supplies and then bioconcentrate in animal and human fatty tissues. To a greater degree than PCBs, these solvents are contaminants of concern on numerous Superfund sites.

Styrene

Styrene is used to manufacture plastics, rubber, and resins. Exposure to styrene may damage the central nervous system, resulting in *dysphoria* among workers in these industries.

Vinyl chloride

We use *polyvinyl chloride (PVC) pipes* made from vinyl chloride to transmit water throughout residences and places of work. We sit on plastic seat covers made from resins of vinyl chloride. When vinyl chloride is polymerized in these products, however, the polymers are not hazardous to our health. Unfortunately, the factory workers who work with the unpolymerized vinyl chloride are exposed to a Class A carcinogenic chemical (evidence of cancer in humans) that is also flammable. Workers in vinyl chloride plastics plants have high rates of incidence of liver and bile cancer. Furthermore, bad batches of PVC generate waste sludge. This sludge contains unpolymerized vinyl chloride that is found at many Superfund sites—especially at those that have unlined landfills. The toxic leachate from the sludge, having no barriers (liners) to contain the leachate, can contaminate nearby water supplies.

Other Organic Contaminants of Concern

Toluene

Still used as a solvent, toluene is a component of gasoline (*octane booster*). It is flammable, but less toxic than benzene.

Acetone

Acetone, like toluene, is still used as a solvent, especially in chemical laboratories and in cosmetics. It is highly volatile, so it is a common laboratory contaminant. In testing for volatile organic compounds, acetone may show up in the analysis even though its only source may be the laboratory itself.

References

1. Girard, J. 2010. *Principles of Environmental Chemistry.* Sudbury, MA: Jones and Bartlett.
2. Friis, R. H. 2012. *Essentials of Environmental Health.* Sudbury, MA: Jones and Bartlett.
3. Hawkins, W. R. 2006. *Eat Right—Electrolyte: A Nutritional Guide to Minerals in Our Daily Diet.* Amherst, NY: Prometheus.
4. Yu, S. J. 2008. *The Toxicology and Biochemistry of Insecticides.* Boca Raton, FL: CRC Press.
5. Cobb, A. H. and J. P. H. Reade. 2010. *Herbicides and Plant Physiology.* Ames, IA: Wiley-Blackwell.

19 Nonchemical Hazards

Physical Hazards

There are other sources of man-made environmental risks that do not involve coming in contact with chemical substances and waste. These include radiation (ionizing and nonionizing), noises and vibrations, extreme pressures and temperatures, and accidents (considered predictable and preventable).

Radiation

Radiation[1] is the transference of *electromagnetic energy* that can be conceptualized as either a wave or a beam of particles. Both concepts are valid; a wave is a photon and vice versa. Electromagnetic energy has wave forms that have crests and troughs with the energy being transmitted for a distance in a prescribed unit of time (the *frequency*). The distance between crest to crest or trough to trough is called one *wavelength*. The time it takes for the energy to travel one wavelength is one period.

Electromagnetic energy can travel through a vacuum and some can penetrate through material. The Sun's light energy travels to Earth through the vacuum of space until it impinges on the Earth's atmosphere. Electromagnetic energy can also penetrate materials. It can go through them without disruption (*nonionizing radiation*), like radio waves or cell phone waves passing through objects. It also can penetrate material by changing the energy level of its atoms (*ionizing radiation*), as used in radiation therapy or by mutating genetic material.

Different forms of electromagnetic energy are distinguished on the basis of their wavelengths. Radio/TV and infrared energies have wavelengths longer than those of visible light (lower frequencies). Ultraviolet and x-ray energies have wavelengths shorter than those of visible light (higher frequencies).

Table 19-2 summarizes the salient features of electromagnetic energy:

Extraterrestrial

Primary *cosmic rays* originate from outer space, mostly coming to Earth from our Sun and other stars in the Milky Way (our galaxy). Upon hitting the Earth's atmosphere they produce secondary cosmic rays. They have the ability to penetrate and cross the human body easily. If one takes an airplane flight or climbs on land to high elevations, the exposure to radiation increases.

Terrestrial (Natural)

Radiation also emanates naturally from rock formations containing varying amounts of uranium. Uranium is a common element in the Earth's crust. Ores of uranium are uraninite and pitchblende, and are found in large amounts in North America, Africa, and Australia. Nearly all plants, animals, and aquifers contain tiny amounts of uranium.

The natural spontaneous decay of uranium and radium emits radioactive atomic particles and produces radioactive *radon gas* (as a daughter product), which can seep into houses through cracks in foundations. Radon is inert, colorless, and extremely toxic. Inhalation of the radon gas can bombard lung tissue with carcinogenic (Class A) alpha particles (each having two protons and two neutrons).

Table 19-2 | Salient Features of Electromagnetic Energy

Conceptual	Energy Movement	Energy Transfer	Possible Effects	Examples
Photons or waves	Bundles of energy or waves flow through a vacuum or penetrate matter	Nonionizing radiation	No changes to matter penetrated	Radio waves, cell phone waves, visible light
		Ionizing radiation	Possible changes to matter penetrated	UV radiation, gamma rays, x-rays

Terrestrial (Manmade)

Using modern electronics, we have learned to create and control *electromagnetic waves* or photons. Basically, when we move electrons, their motion creates a magnetic field, and electromagnetic waves (or photons) are emitted. In this way, we can create *rad*io and television waves, microwaves, x-rays, etc.

Extremely low-frequency (ELF) radiation can be generated from high-tension power transmission lines, wiring in walls of buildings, and some electrical appliances. Even cell phones generate ELF radiation. The danger of using cell phones is not from ELF radiation, but rather from driver distraction while operating a motor vehicle.

Radioactivity

Using nuclear technology, we can cause the nucleus of certain elements to emit radiation. Energy in the nucleus of atoms can be released by the splitting (*fission*) of atoms, actually a process of bombarding certain atoms with subatomic particles and making them unstable. Unstable atoms can do three things: (1) lose protons and neutrons and change into other elements or isotopes of the same element, (2) release subatomic particles, and (3) release large amounts of energy. Emitted subatomic particles can in turn bombard other atoms and make them unstable. The repeated process is quick and, with a critical mass of radioactive material, can form a *chain reaction*.

Radiation Hazards

Radioactivity that might pose an environmental health risk depends on the amount of radioactivity produced, receptor exposure, and the absorbed dosage. This has been measured by four parameters: *curie* (Ci), *rad*, *rem*, and *roentgen* (R). The curie, named after Marie Curie, the discoverer of *radium*, is the amount of radioactivity in a sample of material. The roentgen is a unit of exposure from x-rays or gamma rays. The rem is a measure of dose deposited in body tissue, averaged over the mass of the tissue of interest. The rem replaced the rad, which is defined as the dose irrespective of the mass of tissue irradiated.

Exposure to radiation is dependent on the time of exposure, the distance from the source, and the rate of energy emission from the radioactive material. At low levels, the radiation can burn tissue and cause radiation sickness (e.g., nausea, weakness, loss of hair). At high levels, the radiation can produce fatal injuries. Radiation can also cause cancer and changes in DNA.

Uses of Radiation

Because sources of radiation have the potential to release great quantities of energy and to penetrate solid objects and human bodies, they have many industrial, medical, and military uses.

As just a few examples, radioactive materials can be used to generate electric power, kill cancer cells, change genetic material, manufacture nuclear weapons, and study biological processes.

Noise

Two of our very important abilities involve sensing and interpreting sounds and sights emanating from outside our bodies. Sound is perceived mostly by our ears (also from vibrations felt by our sense of touch). We use our ears to hear conversations, radio and television programs, announcements, news, lectures, music, warning sirens, and signals from our technological devices (e.g., alarm clocks, telephones, timers). Our environment determines what kinds of sounds we hear, including the magnitude (volume) and pitch (*frequency*) of those sounds. Threats resulting in hearing loss are mostly from the sounds we hear, although ear infections may also be a problem. Sounds that range from annoying or actually causing harm are collectively called *noise*.

Physics of Sound

Sound is a vibrating wave that propagates to our ears through the air from its source.[2] Upon arrival at our ears, the vibrating air enters our outer ear canals and pulses air in the canals to change the sound pressure on the outside of the ear drums. The ear drum is an elastic membrane called a *tympanum*. In response to the pressure differences on either side of the ear drums, the membranes, in turn, move small bones in the middle ears. The bones transmit their movements to nerve endings in the inner ears, where these movements are converted to electrical nerve impulses that are sent to the brain for auditory processing. We have two ears that receive sounds from the same source at slightly different times, allowing us to interpret the direction of the source.

Voila, we hear the sound. We interpret the sound. Was the sound loud or soft? Was the sound high-pitched or low-pitched? Was the sound coming from the right or the left? Were the combinations of sounds pleasant or just noise?

The transmission of sounds is in the form of cycling waves that have *amplitude* (loudness or intensity), *frequency* (pitch), and *wavelength* (distance of wave travel in each cycle). Intensity, *sound pressure level (SPL)*, is measured in *decibels (dB)* at the location of the receptor. In a rock band, the amplifier ("amp") intensifies the sound. Whatever the output intensity of the amp, the sound pressure level depends inversely on the distance the receptor is from the amp. Frequency is the number of times that the wave cycles per unit of time. The higher the frequency, the higher the pitch of the sound. A soprano sings notes of higher pitch than a baritone. The wavelength of the sound depends on the frequency and the velocity of the wave. As the velocity is determined by the medium through which the sound wave travels (in this case the air at a fixed temperature), the wavelength is a function of the frequency and can be ignored in our discussion.

Nuisance or Health Threat?

Noise[3] is an unwanted sound. It may be unwanted because it interferes with other sounds we are interested in hearing. Unwanted sounds may drown out the sounds we want to hear. In a crowded restaurant, the background noise may prevent us from hearing the conversations at our own table. Nearby conversations or cell phone usage in a movie theater or at a classical concert may distract us from hearing and enjoying the performance. Our alarm clock may awaken us from a pleasant dream while asleep. Footfalls we hear from the floor above may disturb our concentration. All these noises, unwanted sounds, may cause us stress or depression, but they may not be a direct health threat to our sense of hearing.

Of immediate concern to our hearing health is the sound pressure level. Very loud sounds have large differences in pressure on either side of our ear drums. The scale of SPL ranges from 0 dB (threshold for human hearing) to 194 dB (maximum average air pressure in our environment). In our quiet homes, we may enjoy an SPL of 30 to 50 dB. If someone is talking, this may go up to 60 dB. If someone is operating a home appliance like a vacuum cleaner, we might experience an SPL of 70+ dB. If we stepped outside to hear curbside traffic noises, the SPL level might be 80 to 90 dB. Seated at a rock concert 1 m from a speaker, you might exceed 100 dB. If you took down some trees with a chain saw, the SPL might reach 110 dB. Airport tarmac workers wear ear protection because levels might be greater than 140 dB. Discomfort is experienced at 120 dB and actual pain at 130 dB.

Damage to the Ear

The expression "no pain, no gain" should not be followed when it comes to hearing. In fact, a 70-dB SPL over an 8-hour period or 120 dB for a split second, although not in the range of pain, can cause ear damage. Greater distance from the source reduces the chance of ear damage. Consequently, carefully pretest the volume on devices with ear buds prior to insertion into the ears. Buds transmit sounds directly to the ear canals.

Ear protection should be worn when you use equipment that produces harmful noises. Such equipment includes *lawnmowers (90 dB)*, *jackhammers (130 dB)*, and firearms (140 dB). Ear protection must be worn by the person being tested in a *magnetic resonance imager (MRI)*. SPLs in an MRI can range from about 100 to 130 dB, depending on the equipment used.

According to *Occupational Safety and Health Administration (OSHA)*, the *permissible exposure level (PEL)* for a worker's exposure to noise in an 8-hour day, 5 days per week, is 90 dB. The time of exposure allowed is decreased as the sound level increases. For example, a sound level of 95 and 100 dB shortens the exposure daily time to 4 hours and 2 hours, respectively.

Hearing loss can affect different parts of the ear. *Conductive hearing loss* is from damage to the small bones in the middle ear. *Sensorineural hearing loss* results from damage to the inner ear associated with transmission of nerve impulses.

Audiograms should be administered to test for loss of hearing at various frequencies, when hearing loss or potential hearing loss is suspected. When a receptor anticipates being exposed to loud noises, hearing protection, such as ear plugs or earmuffs, should be used.

Other Adverse Health Effects from Noise

Aside from damage to the ear and hearing loss, noise can adversely affect our health and safety in other ways. Noise can induce stress, distract us during activities that require our undivided attention, and disturb sleep. Driving your vehicle with the car radio blasting or wearing earphones prevents you from concentrating on your driving and from hearing other vehicles' horns, sirens, and train whistles and bells at railroad crossings.

Geological Hazards

Geological hazards are topics discussed by geologists in an Earth science subdiscipline called "*environmental geology*." Most of these topics deal with environmental safety, such as hazards from earthquakes, tsunami, volcanism (**Figure 19-1**), active faulting of the rock strata, subsidence due to mining or fluid extraction, floods, severe weather events, landslides, and radon gas.[4] A relatively new field of geology has emerged called *medical geology*, which is concerned with health effects related to geological media, such as health impacts from airborne particulates (respiratory diseases) and minerals in drinking water (crystal deposition diseases), minerals as medicine or in drug delivery, global climate change, and heavy metal toxicity.

FIGURE 19-1. Stromboli (Italy) Volcano (active) with villages at base.

Biological Hazards

We have survived to the present time, in part, because our ancestors through their child-bearing years developed immunities to diseases, were not exposed to pathogens, had immunizations, practiced healthy lifestyles, lived in areas with good sanitation, and had adequate medical care when needed.[5] Such conditions lead to sustained health and allowed them to pass on genetic-based survival traits and socially-based good health practices.

Biological-based health threats still exist. Threats may result from direct or indirect interaction with other organisms. This can result from *parasitism* (e.g., mosquitos, ticks, chiggers, worms), defensive actions (e.g., snake bites, bee/wasp stings, bear or raccoon attacks), incidental contact (e.g., poison ivy, pet dander, pollen, contagious humans and some animals).

Exposure to pathogens or from the toxins that they excrete can be initiated by the ingestion of contaminated food and water. Once in the body, pathogens or the toxins they excrete can attack our cells. When exposure is initiated by injection from bites from organisms that are not infected themselves but nevertheless carry the disease, the carrier is called a *bio-vector*. All of these threats can cause disease or severe allergic reactions in the receptor.

Workplace Hazards

Luckily, the workplace environment is now monitored for environmental hazards to provide safeguards against physical or ergonomic injuries, heat or cold stress, chemical exposure, and other hazards.[6] A greater degree of risk is tolerated more in the workplace than in other environments because of the use of safety plans, protective equipment, risk communication, and exposure restrictions for sensitive segments of the workforce (physically challenged, elderly, youth, and pregnant women). It is required (by OSHA and some state regulations) that places of work provide information on hazardous substances in use through *Material Safety Data Sheets (MSDS)*. These sheets give information on the substance, exposure limits, characteristics, physical and health hazards, routes of entry into the body, methods to control exposure, precautions for safe handling and use, and emergency contact numbers for additional information. Workers are entitled to know about the hazards they may encounter on the job (right-to-know laws); however, a problem may be encountered when employers want to withhold information on the grounds of protecting proprietary business information.

Sick building syndrome (SBS) is a condition where the cause of indoor sickness complaints cannot be determined—once determined, the situation is called *building-related illness (BRI)*.

The *National Institute for Occupational Safety and Health (NIOSH)* conducts research on occupational disease and injury, investigates (when requested) hazardous working conditions, communicates information on prevention of workplace disease, injury, and disabilities, and provides health and safety training. To prevent health and safety problems on the job, employers are expected to provide engineering controls, to use safe materials and workplace practices, and to provide personal protective equipment and clothing when necessary. In addition, employers can conduct medical surveillance, identify problems or symptoms early, modify work programs and provide medical care and first aid.

Hazardous workplace environments require workers and supervisors to undergo rigorous training and abide by health and safety plans. Supervisors at hazardous workplace sites must keep records of employees' health profiles, provide employee training, utilize health and safety plans to make employees aware of hazards, conduct safety meetings, have protective equipment and clothing on hand, control the spread of hazardous materials by decontamination protocols, and monitor and control ambient temperatures, wind direction, humidity, and entry and exiting of restricted areas. In spite of safeguards, incidents occur, and employers and employees must respond according to established procedures.

References

1. Friis, R. H. 2012. *Essentials of Environmental Health.* Sudbury, MA: Jones and Bartlett.
2. Berg, R. E. and D. G. Stork. 2005. *The Physics of Sound.* Upper Saddle River, NJ: Pearson Prentice Hall.
3. Wilson, A., comp. 1988. *Noise Pollution.* Washington, DC: Science and Technology Division, Library of Congress.
4. Keller, E. A. 2012. *Introduction to Environmental Geology.* Upper Saddle River, NJ: Prentice Hall.
5. Grady, S. M. and J. Tabak. 2006. *Biohazards: Humanity's Battle with Infectious Disease.* New York: Facts on File.
6. Rose, V. E. and B. Cohrssen, eds. 2011. *Patty's Industrial Hygiene.* Hoboken, NJ: Wiley.

Population Growth

Image © leungchopan, 2014. Used under license from Shutterstock, Inc.

Is Population Growth Sustainable?

In 2011, the world population reached 7 billion people. Two and a half billion of those are in China and India. The world population at the beginning of the Common Era (year 1 AD) was only about 200 million. So we have doubled in population almost five times in just over 2,000 years (years 1 to 1400, 1400 to 1800, 1800 to 1927, 1927 to 1975, and 1975 to ?, est.). Each succeeding doubling took much less time (1400 years, 400 years, 127 years, 48 years, and ? years).

The *doubling time*[1] is calculated by dividing the number 70 by the annual rate of population increase. So if the *growth rate* is 2 percent, the doubling time is 35 years (70÷2). The *growth rate* is determined by the difference in annual *birth rate* and *death rate* (per 1,000), converted into percent (per 100). Therefore, if the annual birth rate is +30 and the annual death rate is −20, the total increase is 10 per 1,000 people, or 1 percent. If we divide 1 into 70, we get 70 years as the time it would take to double this population.

Some countries in Asia, Africa, and Latin America have high growth rates, whereas some countries in Western Europe have a low or negative growth rate. Some of those same countries with high growth rates have a high percentage of youth (fertile or soon to be fertile), indicating a trend for higher growth rates in the future.

We assume that Earth has a finite amount of resources. Therefore, if our population increases, we will run out of resources more quickly and may experience dire consequences concerning the sustainability of the human race (**Figure 20-1**). Is this assumption correct?

FIGURE 20-1. Resource depletion gauge

Paul Ehrlich and his wife, Anne,[2] in the late 1960s insisted that this assumption was indeed correct. Although their predictions about widespread starvation may have been true in third-world countries, in succeeding decades following the publication of their book, their alarming ideas have proven inaccurate with respect to the majority of people in industrialized and technologically advanced countries. Lack of resources may be the result, in part, of inequitable distribution of goods rather than from true worldwide shortages. The result is a gap between wealthy and poor (haves vs. have-nots, or "have-nots vs. have yachts").

The Ehrlichs' views supported the theories of a scholar whose observations were made over a century and a half before. Thomas Malthus ("An Essay on the Principle of Population, 1798–1826") observed that births are offset by population declines from famine and disease. Indeed, populations have also been kept in check by wars.

Opposition to the Ehrlichs' views of a population doomsday was espoused by the late economist Julian Simon.[3] He felt that the most valuable resource was the intellect of humans, who could find ways of surviving and prospering in greater numbers with available resources. To support his views, Simon placed a bet with Paul Ehrlich about the cost of metals over the succeeding 10-year period. They picked five metals. Simon said the prices of these metals would decrease over a decade, and Ehrlich said they would increase. Ehrlich's price increases were predicated on the assumption that the supply of those metals would be depleted. Lower supply equals higher demand, and higher demand results in higher prices. Who won the bet? Simon did. All the metals had a lower price 10 years later. Why? We have learned to extract those metals more cheaply, recycle them, and find substitutes. As a result, the demand for those metals decreased, driving down their prices.

Malthus's population checks are not working with respect to food resources. We keep producing more food on less available farmland. Malthus's population checks are not working with respect to diseases, as medical technology has found ways of preventing pandemics, curing diseases in humans during and prior to childbearing years, and increasing life expectancy. We are still working on preventing wars, which kill soldiers (generally of childbearing age) and the general population (collateral damages).

It is in our general interest to prevent disease, famine, and war. It is also in our general interest to limit population increases to match the demands on our resources. Population checks, as needed, should not rely on famine, disease, and war. Rather, this aim should be accomplished by limiting the size of our families, preferably voluntarily.

Measuring Growth

The *populations* of any species, humans included, is the total number of individuals of that species. When you enter a town on the highway, you might see a road sign indicating the population (of humans) in that town according to the latest census. This number does not include other species (of birds, insects, raccoons, etc.), just humans. The populations of any one species of insect might be in the order of millions, and the number of one species of bacteria might be in the order of billions in that town. But humans in each area are doing the counting, so they only count themselves.

Let us define some terms about the characteristics of populations.[4] The birth and death rates are, respectively, the number of one species born or who died per thousand individuals in 1 year's time. The growth would be the difference of births to deaths, which could be positive (growing population) or negative (declining population). The growth rate, however, is measured as a percentage.

The number added to the population through reproduction (not *immigration*) is the *natality*. The number subtracted from the population through death (not *emigration*) is the *mortality*. To sustain a population, natality should equal or exceed mortality. Therefore, *sustainable populations* should rely on the reproductive capacity (*biotic potential*) of those individuals of reproductive age. The distribution of young to old and females to males affects the growth rate. In populations having a high percentage of individuals beyond reproductive years, the longevity of the older individuals is the determining factor for the population growth rate of the species, as they cannot directly affect the natality. "Directly" is mentioned, because older individuals can encourage or discourage the young to produce more or fewer offspring. For humans, the influence of the elderly (politically, economically, or through religious practices) can encourage marriage, dictate family size, and discourage contraception, to name a few possible impacts.

The *population density* is the concentration of individuals in a given area (**Figure 20-2**). This statistic may be misleading, as humans may choose to live in crowded areas close to water sources, fertile lands, or jobs in industry, leaving other areas sparsely populated. Density is determined not only by natality and mortality, but also by the dispersal of the population by *emigration* and *immigration*. Population density may affect health as crowded communities put a strain on sanitation services and increase contagion of communicable diseases.

FIGURE 20-2. High population density in Rio de Janeiro

Dangers of Overpopulation

In any one area, the maximum number of a species that can be sustained using available resources is called its *carrying capacity*. Most nonhuman populations are strictly limited by carrying capacity, which they cannot change. If the species has a population greater than the carrying capacity, individuals die off—populations "crash"—until the population is less than the carrying capacity. Then the population can continue to survive and rebound in numbers. Each time the carrying capacity is exceeded, the population crashes again, and so on.

Humans, however, can raise the carrying capacity by technological improvements. We can grow more food using improved farming practices. We can transport food greater distances to where it is needed in a shorter time. We can improve the sanitation of communities. We can provide a cleaner environment. We can provide medical care to greater numbers. We can immunize populations against diseases. We can develop new cures and treatments for ailments. We can reduce human health risk factors by improving the lifestyles we lead.

How high can the human carrying capacity be raised? We do not know. Julian Simon would have said that the potential to increase the carrying capacity of humans is limitless. Skeptics would disagree.[5]

Skeptics would focus on poor people in the developing world. Their limited economy and lack of access to the latest technologies prevent them from raising the carrying capacity as easily as those in the developed world. All of the tools for raising the carrying capacity take funds and know-how. The have-nots, where the population growth rates are also the highest, are the poorest and least educated humans.

Will this lead to population increases in the underdeveloped nations? Sometimes this is true. Families choose to have more children in these societies to produce more wage earners (farm hands, child laborers, or beggars). As one of the authors of this text was working as a consultant for a government agency in an underdeveloped country in Asia, he came upon child beggars. He noticed that some were blind. He asked one of the agency professionals if a disease was the cause of the blindness. They informed him that parents have been known to blind their own children to make them better beggars.

Whether they are hungry, in need of water, or searching for ways to improve their situation, people's frustrations may lead to violent confrontation with their neighbors or, on a larger scale, with wealthier nations. We see this historically as well as today over energy, agricultural lands, and water supplies. In the future, water supplies will become a more important demand and a more important cause for frustrations in the developing world.

Will this lead to decreases in populations? Yes. Frustration over resources can lead to conflicts. Conflicts can lead to wars and terrorism, which increase the risks of disease, death, and disruption of human services. All this would adversely affect medical care, sanitation, education, and environmental protection.

Striving for a Sustainable Population

To solve the problems of overpopulation, we must either limit population increases to the carrying capacity of our current resource potential or increase the carrying capacities of all nations, rich or poor, to accommodate ever-increasing populations worldwide. The difficulty arises from the fact that population increases are greatest in the segment of the population where it is most difficult to increase the carrying capacity. Food is the most important resource lacking in the have-not populations of humans.

The *World Food Programme* estimates that there are 925 million people who do not get enough to eat to lead normal, active lives. This situation is called *undernourishment*. They do not get enough calories to meet minimum physiological needs. Most undernourished people live in Africa south of the Sahara, South East Asia, the Middle East, Central America, and northern South America. Lack of the proper diet can lead to *malnutrition*, or an inadequate intake of a balance of macro- and micronutrients. Malnutrition prevents the body from fighting off infections and diseases.

Famine, as defined by the United Nations, is a situation where 30 percent of children are acutely malnourished, 20 percent of the population is without food (<2,100 kcal/day), and deaths are running at 2 per 10,000 adults or 4 per 10,000 children every day.[6]

Humanitarian food relief supplies sent to poorer nations too often do not get to the people who need them most. This is in part a political problem, which exacerbates the problem of world hunger. Food shipments are also hijacked during transit to the intended recipients.

References

1. Johnson, A. T. 2011. *Biology for Engineers.* Boca Raton, FL: CRC Press.
2. Ehrlich, P. R. 1968. *The Population Bomb.* New York: Ballantine Books.
3. Simon, J. L. 1998. *The Ultimate Resource 2.* Princeton, NJ: Princeton University Press.
4. Bellamy, R. 2011. *Population Growth.* Mankato, MN: Amicus.
5. Hohm, C. F. and L. J. Jones, eds. 1995. *Population: Opposing Viewpoints.* San Diego, CA: Greenhaven Press.
6. Jacobs, A. and DuWayne, P. 2012. *Environmental Science—Health Impacts.* Dubuque, IA: Kendall Hunt Publishing.

21 Poverty

Capitalism vs. Socialism

The cause of *poverty* is twofold: (1) not enough resources to go around and (2) disparity between the haves and the have-nots.[1] In the twentieth century, human societies experimented with technology and economic/political systems in trying to solve this twofold problem. For the problem of not having enough to go around, technology stepped in to increase the carrying capacity to sustain a rapidly growing population. Revolutions in agriculture, industry, and engineering increased supplies of food, created high-tech machines, increased efficiency in manufacturing (assembly lines, division of labor), and streamlined communication and transportation. Part one seemed to be solved for the industrialized countries . . . for now, but the developing countries have not fully participated in this increase in carrying capacity.

FIGURE 21-1. Shantytown, Vietnam

The solution to part two has been more difficult, namely, reducing the gulf between the rich and poor—providing economic justice for all. Many rich people and their political supporters have espoused that there are not enough resources to go around. The "pie" has a limited size or would not grow if resources were shared. They believe that the sharing of resources would make everyone live at or near the poverty level, and there would not be any incentive to sustain or grow existing wealth[2] **(Figure 21-1)**.

Many hardworking but poor and their political supporters experienced a dwindling economic return for their labors. Extreme political systems were maintained, modified, or established at the beginning of the twentieth century to hopefully satisfy the rich and the poor, with mixed results. Untamed *capitalism* was modified (*social welfare*) or replaced (*communism*). Countries still operating under capitalistic economies developed perks for workers, e.g., labor/management decision making (unions), retirement plans, reduced work week/hours, workers' health and accident insurance. This did not satisfy some of the people at the very top of the economic ladder, who maintained that we should privatize or keep private the control of workers' perks. Consequently, subsequent employer practices and legislation shifted jobs overseas to developing countries where salaries were much lower, displacing workers in more highly paid jobs in developed countries.

In the twentieth century, two populous countries had political revolutions and established communist governments that set up anticapitalist economic systems. Both the Soviet Union (and its satellite neighbors in Eastern Europe) and mainland China set up communist systems in political dictatorships espousing the merits of a social economy that provided prosperity for all. What resulted, however, were economic gains for the political rulers and resource shortages (e.g., bread-lines) for most of the citizens. Smaller countries that adopted this system included Cuba, North Korea, and North Vietnam.

As the twenty-first century began, replanting the seeds of some aspects of capitalism has occurred in most of the communist societies mentioned. Although no other populous countries have since adopted strict communist governments, a socially modified form of capitalism in industrialized political democracies has been maintained. In fact, many workers' perks have increased. Some western European countries mandate 5-week vacations with expense allowances, generous maternity and new-parent leaves with pay for both women and their spouses, early retirement with government pensions, and job security.

Environmental Pollution Perpetuates Poverty

How does environmental pollution perpetuate poverty? What makes us poor is not just a shortage of affordable resources, but also hidden costs and consequences from pollution. Many companies that generate polluting wastes deny responsibility for causing disease and shift cleanup and medical costs to society in general. The poor are the least likely to afford these medical costs or have political clout to fight the polluters. Furthermore, urban sources of pollution (e.g., from factories, refineries, mills) are usually located on real estate zoned for industry, but are adjacent to poor neighborhoods. Housing costs are low in these neighborhoods, where exposure to pollution is greater, so the poor cannot afford to move away either. Consequently, poverty reduces our ability to maintain a clean and healthful environment. If a healthful environment is not sustainable, then poverty instead becomes sustainable.

Food Surpluses—Will this Help?

In wealthy countries, increased agricultural production is keeping pace with the current population. Productive agricultural countries, like the United States, have food surpluses, some of which has been donated to poorer countries. However, food distribution is not getting needed nutrition to many segments of poorer societies because of trade barriers and corruption. Some of our capacity to generate surpluses is being diminished. Crops that could be used for food, like corn, are being used instead to produce sustainable fuel (ethanol and methane). Farm acreage adjacent to urban and suburban areas is being transformed for other land uses (residential and commercial).[3]

References

1. Anand, S., P. Segal, and J. E. Stiglitz, eds. 2010. *Debates on the Measurement of Global Poverty.* New York: Oxford University Press.
2. Cornia, G. A. 2004. *Inequality Growth and Poverty in an Era of Liberalization and Globalization.* Oxford: Oxford University Press.
3. Sen, A. 1983. *Poverty and Famines: An Essay on Entitlement and Deprivation.* Oxford: Oxford University Press.

Climate Change

Image © Sam72, 2014. Used under license from Shutterstock, Inc.

Climate change, or specifically *global warming*[1], is a popular topic for political discussion. Since ex-Vice President Al Gore published his book and movie documentary, "An Inconvenient Truth," politicians have lined up on either side of the debate. Some deny that climates are changing at all. Some maintain that the climate is actually getting colder. Others insist that the climate is being degraded by our rapid generation of carbon dioxide (and other greenhouse gases). Still others on both sides of the argument point to the increased storms and other catastrophic weather events that are flashed on television and Internet notices to alert and update us on the current or next storm. Antichange forces point to cold spells during the winter to refute that the climate is getting warmer. Prochange forces show how the rate of greenhouse gas emissions, a cause of global warming, has skyrocketed. Let us look at the evidence.

Causes of Change

Our dynamic Earth is a changing planet, and climate is no exception. The causes (not one) are as follows: fluctuations of solar energy output, changes in Earth's orbit, wobble of the Earth on its axis, changes in the circulation of the atmosphere and oceans, *albedo* of the Earth to reflect solar radiation back into space, and the presence in our atmosphere of *particulates* (*dust*) and *greenhouse gases*.

The rate of solar output varies, correlating with the appearance of sun spots and solar flares. This changes the rate of solar heat input.

The orbit of the Earth around the Sun is roughly elliptical, but the actual dimensions of the ellipse vary from year to year. This affects the distance of the Earth to the Sun.

The axis of the Earth on which it rotates is not at a constant angle to the plane of orbit; it actually wobbles. The wobbling motion affects the angle at which sunlight distributes its heat energy onto the Earth's surface.

Atmospheric and oceanic circulations affect each other. The atmosphere circulates air and moisture, and the oceans circulate water currents. Both participate in the hydrologic cycle. Moisture is evaporated from the sea into the atmosphere, which returns to the sea directly from rain and snow and indirectly from the rain and snow running off the land into the sea. Both exchange and distribute heat energy. Both contribute to weather events.

Snowfields and glaciers can reflect solar heat back into space (*albedo effect*). The more snow and ice cover, the greater the albedo producing cooler temperatures on Earth. As snow and ice cover diminishes, more heat can be absorbed on Earth.

Natural phenomenon (e.g., volcanic activity, sea spray, many forest fires, organism respiration, weathering of rock, vegetation decay) and human activity (e.g., combustion engine exhausts, landfill gases, furnaces at mills and power plants, mining and construction, ore and petroleum processing) add to particulates and gases in the atmosphere. Greenhouse gases (*carbon dioxide*, methane, *chlorofluorocarbons (CFCs)*, and *water vapor*) help to retain heat from escaping our atmosphere. Particulates, in reverse, causing overcast cloudiness keep heat from penetrating the atmosphere to the surface.

Which of these causes has the greatest effect on climate change? Perhaps the history of change can help us to answer this question.

History of Change

In the last 2 million years (and probably before that), the Earth has experienced dramatic climate changes, especially with regard to *glaciation*.[2] The Earth was covered by *continental ice sheets* extending from the polar regions and by *valley glaciers* extending from the highest mountains. More than four glaciations occurred in the last 2 million years with long periods of warmer climate (even warmer than today), so-called *interglacials*. We are currently coming out of a major ice age that peaked at about 30,000 years BP (before present = 1950 AD) and waned about 10,000 years BP. This was followed by a warm period (the *hypsithermal interval*—9,000 to 5,000 years BP). Then in the 16, 17, and 18 hundreds cooling trends interrupted this postglacial warming period, when glaciers expanded again (so-called *Little Ice Age*). These changes occurred without the influence of huge contributions of industrial greenhouse gasses. They resulted from the other causes of climate change.

FIGURE 22-1. Edge of Jakobshavn Glacier (Greenland) calving icebergs into the sea

During each glaciation, the increased cover of reflective ice and snow increased the Earth's albedo and increased the cooling process. Also, excess water on land adding to the supply of ice and snow reduced sea level and increased the acreage of continental land areas. The seas became saltier. Conversely, during each interglacial period the albedo decreased and warming of the surface occurred. Also, sea level rose again inundating continental margins with sea water. Melting also reduced ocean salinity adjacent to the ice sheets and affected ocean currents **(Figure 22-1)**. The cooling and heating was cyclic in nature. What triggered the change?

Two geologists/geophysicists at Lamont-Dougherty Laboratories at Columbia University, Drs. Maurice Ewing and William Donn, theorized that warming caused the onset of ice ages.[3] This sounds like an oxymoron—how can warm cause a climate change that produces a cycle of colder weather?

To understand this, one must understand a modern-day phenomenon called *lake effect snows*. For example, when the Great Lakes are not frozen in the early winter, moisture that is evaporated from the lakes is swept onto the downwind lake shores by cold wintry winds and dumps heavy snowfalls. After the lakes are frozen, the lake effect snows diminish. This process can be applied to the Arctic Ocean.

The Arctic Ocean covers a large area between northern Alaska and Canada and the northern border of Asia and Europe. Normally, the Arctic Ocean is frozen over in the early winter and remains frozen until the early summer. Much ice remains through the summer and the freezing process begins again in late autumn. The downwind shoreline of the Arctic Ocean is very cold and windy in winter, but has relatively low snowfall. Descending air currents at the north polar region also allow the air to retain much of its moisture.

The effect that Ewing and Donn proposed, here called the *Ewing–Donn theory*, is an ice-free Arctic Ocean in a warmer period. The resulting increase in winter snowfall and ice cover of land in the Arctic Circle increased the Earth's albedo and cooled the climate and encouraged glacier growth. This process reversed itself when the supply of moist air from the Arctic Ocean was hindered by the reestablishment of the ice cover of the Arctic Ocean and reduced snowfall on its shores. The reduced albedo then helped to initiate the next warming trend.

This theory does not negate the causes explained above. It emphasizes geologic conditions on Earth, especially changes in albedo and ocean–atmosphere interaction, that explain the major cycles of glacial and interglacial climates. All the causes explained above are still valid.

The current debate, however, concerns the effect of human air pollution on climate change. The dramatic increase in greenhouse gases emitted by humans into the atmosphere in the last century cannot be ignored. Humans are affecting climate change. What is uncertain is: how much change will there be in the foreseeable future?

What Will the Future Bring?

The *evidence of* climate change is being documented. We are experiencing rises in sea level or lowering of global temperatures, albeit minor. Most of the temperature rise is in the arctic zones or in the high mountains. The Arctic Ocean is progressively becoming more ice-free with each passing summer. The Antarctic ice sheet is losing mass along its edges. Valley glaciers in high mountains are disappearing or their fronts are melting back (*retreating*). We are experiencing severe weather conditions, but not unlike conditions in the recent past. What has changed recently are weather alerts, updated often, that we can get on 24-hour televised weather channels, on our smartphones and tablets, and through social media. Is the climate getting worse fast, or are weather alerts creating an appearance of impending doom? Are climate change deniers convincing people that a cold snap in winter disproves a warming trend?

It is hard to believe that we will significantly change our habits of greenhouse gas emissions to significantly reduce those emissions. Would the cutoff of all combustion emissions of carbon dioxide remove the threat of global warming? Would a warming period trigger another ice age, Little Ice Age, or cooling interval as the Ewing–Donn theory suggests?

The most drastic outcome of a *full-interglacial warming period* or a new hypsithermal interval is massive glacier melting leading to an approximate 300-foot rise in sea level, which would cause coastal cities to become completely inundated. A slow rise may allow for incremental evacuations and sea walls, dikes, and levees to be put in place to protect some areas from flooding. Massive melting would take time at current rates of annual temperature increases.

Other outcomes of a global warming might include a disruption in the populations of creatures that are tied to changes in temperatures. Wildlife habitats would change, ecosystems would rebalance, and biomes would shift geographically. Tropical biomes would displace some temperate biomes; temperate biomes would displace some boreal biomes; and boreal biomes would displace some tundra. Insect populations that favor warmer climates would increase, with vector species spreading disease over longer periods of time and wider geographic areas. On the bright side, agriculture might experience a longer growing season. Nevertheless, if these changes are gradual, Mother Nature would adjust along with the climate changes.

Other aspects of our environment that are not tied to climate would not change. Biorhythms that are tied to daily (lengths of daylight and nighttime hours), tidal, or annual cycles would not change. Some seasonal changes are actually tied to daily cycles rather than to temperature changes.

People believing that reducing our carbon footprint would avoid climate change may give too much credit (or blame) to the power of the human race, even with its technology.

Can we boast that someday we might need to increase greenhouse gas emissions to avert another ice age? The environment, however, is very complex. Mother Nature, also, is very powerful. Natural forces that we learn about in environmental science would counteract what methods we employ to either change or maintain the status quo.

References

1. Haley, J. *Global Warming: Opposing Viewpoints.* San Diego, CA: Greenhaven Press.
2. Wilson, R. C. L., S. A. Drury, and J. A. Chapman. *The Great Ice Age: Climate Change and Life.* New York: Routledge.
3. Ewing, M. and W. L. Donn. *Theory of Ice Ages, Science 123* no. 3207 (1956): 1061–66.

UNIT 6

Correcting Environmental Mistakes

Emergency Preparedness and Response

CATASTROPHES

Cataclysmic events that degrade health and the environment can be caused by humans or can result from events that are not initiated by human actions. The latter events we call *natural disasters*. Natural disasters result from geological, meteorological, and biological causes. They include:

GEOLOGICAL[1]

- Earthquakes
- Tsunamis
- Volcanic eruptions
- Mass movement of surface materials (e.g., landslides, mudflows, avalanches)
- Surface rock displacement (faulting)
- Land subsidence from subsurface solution of rock

METEOROLOGICAL[2]

- Hurricanes/typhoons and other storms
- Tornadoes
- Meteorite impact
- Lightning-ignited forest and range fires
- Monsoons and floods

BIOLOGICAL[3]

- Disease pandemics
- Localized contacts with other organisms (bites, parasites, red tides)

Humans can initiate disasters that degrade health and the environment by their own actions (or inactions).[4] The following are examples:

- Wars
- Terrorism
- Pollution
- Overpopulation leading to hunger and starvation
- Settling in areas that are prone to natural disasters
- Carelessness

Of these natural disasters, some cannot be prevented. Some can be mitigated. For all of them, however, preparations can be made by proactive and reactive actions by public health and safety agencies and by an informed populace. Preparation for disasters always seems like a good idea in their aftermath, but such intent is too late. The damage is already done. What about preparing for the next disaster? Unfortunately, our priorities to be proactive seem to dwindle with the passage of time.

What have we learned in the past 2,000 years? We know more about the causes of natural disasters and the dangers of human-caused disasters. However, ignoring the problems by providing excuses (economics, misguided priorities, and false optimism) can be illustrated time after time. Indeed, one lesson from the past (2,000 years ago) serves as a reminder to continue educating the public on the dangers we face in the future.

Pompeii (79 AD)

A European city in the southern part of the Roman Empire, Pompeii, was a few kilometers from Mt. Somma, a precursor of Mt. Vesuvius.[5] It was a conical mountain with a crater on top that spewed steam from time to time. "Were these vapors from Hades?" some Pompeians might have wondered. The mountain slopes supported flocks of sheep and goats, peacefully grazing on the grasses. A small earthquake occurred in 77 AD that caused minimal damage to the town. Life went on. Public health concerns focused on a plague in areas of the empire to the north. That plague killed 300,000 people, but did not reach Pompeii. Then in 79 AD Mt. Somma erupted and formed Mt. Vesuvius. In doing so, it released clouds of volcanic ash that descended on the town of Pompeii and the surrounding region, which suffocated about 2,000 people. The town structure was not destroyed, only the people in it were. Casts of victims were found intact with their hands over their mouths and noses (**Figures 23-1 and 23-2**). The remains of other victims were found in the nearby town of Herculaneum.

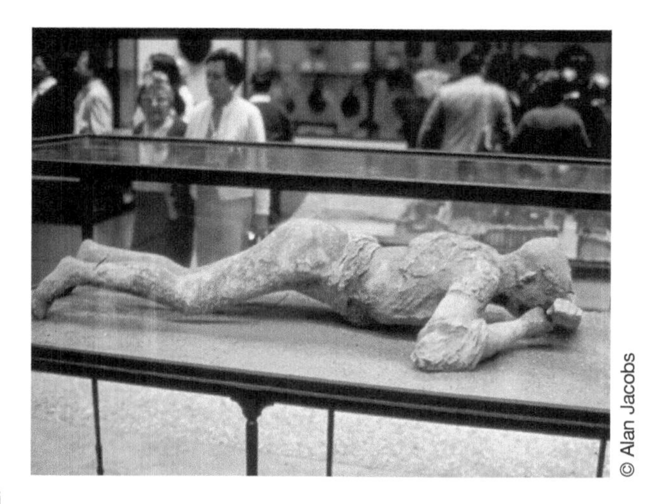

FIGURES 23-1 and 23-2 Casts of Vesuvius victims at Pompeii

These towns were exhumed during the twentieth century as historical tourist sites after centuries of looting. Yet, the towns today appear to be in good structural condition with the inhabitants removed, except for body casts in Pompeii (now Pompei, Italy) and some bones of human skeletons found in Herculaneum (now Ercolano, Italy). As a tourist, you can stroll the streets, view the gouges of chariot wheels in the stone pavements, visit former shops and residences, read the Latin graffiti (political advertisements) on the walls, go to the public forum and arena, and see some of the artifacts in the museums. These towns did not prepare for the disaster. Nowadays, the nearby city of Naples, one of the most populous cities in Italy, has an emergency preparedness plan for the next big eruption. Unfortunately, on a normal day in and around the city, the traffic would not allow evacuation on short notice. We do not learn from the past.

Government Infrastructure for Responding to Emergencies

What do California wildfires, hurricanes, and the terrorist attacks of September 11, 2001, have in common? All of these events, whether natural or created by man, showed a weakness in how we respond to disasters. Our responses showed:

- undefined lines of authority and communication,
- a lack of reliable information to a single source,
- a lack of structure and coordination between agencies, and
- confusion caused by terminology differences between agencies.

All of these incidences typically begin and end in a local area under local jurisdiction. There was a need to develop an effective and efficient coordination across agency and jurisdictional boundaries. These agencies included local, state, federal, and tribal units. On February 28, 2003, President Bush issued Homeland Security Presidential Directive 5 (HSPD-5), "Management of Domestic Incidents," which directed the Secretary of Homeland Security to develop and administer a National Incident Management System (NIMS).

This system allows federal, state, local, and tribal governments along with nongovernment organizations to address the effects of emergencies regardless of size. HSPD-5 requires all federal departments and agencies to adopt and use NIMS.

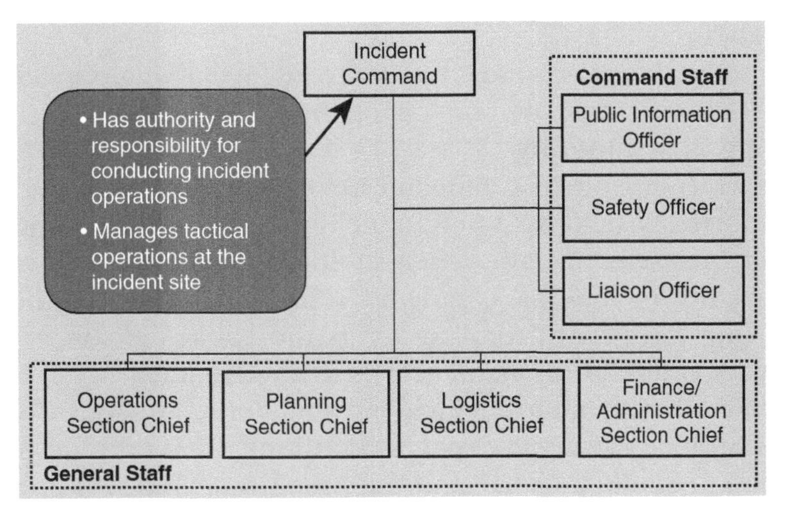

FIGURE 23-3 ICS organization chart (from the Federal Emergency Management Administration, FEMA)

The next step was to choose a command system that would meet the needs of NIMS.

The concept of an incident command system (**Figure 23-3**) had been developed in 1970 when wildfires devastated California. The overall cost of these fires was $18 million per day, and 16 lives were lost. As a result, Congress ordered the U.S. Forest Service to develop a command system to address the communication and coordination problems encountered. A team of emergency services departments was formed and labeled FIRESCOPE (Firefighting Resources of California Organized for Potential Emergencies). FIRESCOPE formally agreed upon a system formally called the Incident Command System (ICS). This system[6] answered the need to manage problems rapidly. ICS was designed to address four basic principles.

1. The system must be organizationally flexible to meet the needs of incidents of any kind and size.
2. Agencies must be able to use the system on a day-to-day basis for routine situations as well as for major emergencies.
3. The system must be sufficiently standard to allow personnel from a variety of agencies and diverse geographic locations to rapidly meld into a common management structure.
4. The system must be cost-effective.

Though the ICS applications were originally designed to respond to disastrous fires, it is noteworthy that the same characteristics are seen in many situations with law enforcement, public health agencies, hazardous material teams, hospitals, and many other agencies and first responders. ICS is based on proven management tools that contribute to the strength and efficiency of the overall system. The system is composed of two basic units: (1) command staff and (2) general staff.

1. ICS Command Staff
   Command staff comprises the *Incident Commander (IC)* and his/her staff. Staff positions are established to assign responsibility for key activities not specifically identified in the general staff functional elements. These positions may include the Public Information Officer (PIO), Safety Officer (SO), and Liaison Officer (LNO), in addition to various others, as required and assigned by the IC.

Incident Commander

- *Single Incident Commander*—most incidents involve a single incident commander. In these incidents, a single person commands the incident response and is the decision-making final authority.
- *Unified Command*—a Unified Command involves two or more individuals sharing the authority normally held by a Single Incident Commander. Unified Command is used on larger incidents, usually when multiple agencies are involved. A Unified Command typically includes a command representative from the main involved agencies and one from that group to act as the spokesperson, though not designated as an IC. A Unified Command acts as a single entity.
- *Area Command*—during multiple-incident situations, an Area Command may be established to provide for ICs at separate locations. Generally, an Area Commander will be assigned (a single person), and the Area Command will operate as a logistical and administrative support. Area Commands usually do not include an Operations function.

Command Staff

- *Safety Officer*—The Safety Officer monitors safety conditions and develops measures for assuring the safety of all assigned personnel.
- *Public Information Officer*—The Public Information Officer (PIO) serves as the conduit for information to and from internal and external stakeholders, including the media or other organizations seeking information directly from the incident or event. While less often discussed, the Public Information Officer is also responsible for ensuring that an incident's command staff is kept apprised of what is being said or reported about an incident. This allows public questions to be addressed and rumors to be managed and ensures that other such public relations issues are not overlooked.
- *Liaison Officer*—A Liaison serves as the primary contact for supporting agencies assisting at an incident.

2. General Staff

   The General Staff includes incident management personnel who represent the major functional elements of the ICS, including the *Operations Section Chief*, *Planning Section Chief*, *Logistics Section Chief*, and *Finance/Administration Section Chief*.

   General Staff

   - Operations Section Chief—The Operations Section Chief is tasked with directing all actions to meet the incident objectives.
   - Planning Section Chief—The Planning Section Chief is tasked with the collection and display of incident information, primarily consisting of the status of all resources and overall status of the incident.
   - Finance/Administration Section Chief—The Finance/Administration Section Chief is tasked with tracing incident-related costs, personnel records, requisitions, and administering procurement contracts required by Logistics.
   - Logistics Section Chief—The Logistics Section Chief is tasked with providing all resources, services, and support required by the incident.

Command staff and general staff must continually interact and share vital information and estimates of the current and future situation and develop recommended courses of action for consideration by the IC.

To address all aspects of the incident, the Planning Section is responsible for developing and documenting an *Incident Action Plan (IAP)*. The IAP includes the overall incident objectives and strategies established by the IC. This planning process is accomplished with productive interaction between jurisdictions, functional agencies, and private organizations. The IAP also addresses tactical objectives and supports activities for one operational period, generally 12 to 24 hours. The IAP also contains provisions for continuous incorporation of "lessons learned" as identified by the Incident Safety Officer or incident management personnel as activities progress.

This system (ICS) has allowed governmental, nonprofit, and private agencies to completely and smoothly address the needs of varying disasters, and it continues to grow in popularity and use. The *Federal Emergency Management Agency (FEMA)* offers a number of free online courses to address training needs.

Mustard Gas Incident in Bari, Italy

You cannot be prepared if you do not know what to prepare for. This was the case near the close of World War II (WWII).

The Allies (the United States, Britain, France, and Russia) were fighting against the Axis (Germany, Italy, and Japan) from 1939 to 1945. The United States did not enter the conflict until after the bombing of Pearl Harbor (Hawaii) in December 1941. In 1943, after the Allies invaded and occupied southern Italy and started pushing northward, many supply ships arrived and docked in the port of Bari, Italy **(Figure 23-4)**, at the "heel" of the Italian peninsula on the Adriatic Sea.[7] The Germans decided to bomb the supply ships in the Bari harbor to destroy fuel for the Fifteenth Air Force, which was attacking Axis strongholds to the north. The attack was also focused on stopping the northward advance of Allied troops from southern Italy.

The Allies knew that the Germans had used mustard gas in warfare during World War I (WWI) (1914–1918). Mustard gas is a chemical agent that causes blisters on exposed skin and lungs, temporary blindness, swollen genitals, and irritated respiratory tissue. The Allies wanted to be ready for a counter-offensive in WWII should the Germans use this chemical again.

Consequently, the United States sent a ship, the *John Harvey*, in convoy from Baltimore to the port of Bari to be held in reserve. It contained two thousand 100-lb mustard gas

FIGURE 23-4 Present-day view of the old port of Bari, Italy

bombs. To maintain secrecy, the harbor command was unaware of the cargo of the *John Harvey*, so unloading of its cargo in the busy harbor was not scheduled immediately. Five days after the *John Harvey* arrived at the port and still waiting to unload, German aircraft bombed the harbor.

The *John Harvey* was set on fire from debris from adjacent ships that had suffered direct hits from the bombing aircraft. The fire caused the *John Harvey* to explode and killed all persons on board. Mustard gas was released onto the oil-slicked water surface, which caused the chemical to spread into the air in clouds of toxic smoke. About 1,000 military and an equal number of civilians were exposed to the chemical. About 800 were admitted to local hospitals. Hospitals did not at first know the reason for the victims' symptoms and allowed them to remain in their chemical-soaked clothing. There were many deaths, but most victims recovered from their exposure to the chemical.

Authorities in the port prior to the bombing and the emergency responders after the bombing could not have prepared for this incident because of the shroud of secrecy around the cargo. Even if the shipment of mustard gas was justified for defense, secrecy prevented a quick unloading of the mustard gas bombs. After the explosion of the *John Harvey*, secrecy prevented timely treatment of the victims. The lesson from this disaster is clear: One must know about the potential hazards to be prepared for those hazards.

Abandoned Sites Needing Immediate Action

Immediate action may be required on abandoned hazardous-waste sites once they are discovered and when the situation threatens immediate harm to human health and safety and the environment. This may include removal of waste, banning water use and providing an alternate water supply, or evacuation of receptors. A complete site cleanup can progress after necessary emergency measures are satisfactorily completed.

References

1. Stambaugh, H. and D. Sensenig. 2008. *Special Report: Fire Department Preparedness for Extreme Weather Emergencies and Natural Disasters.* Emmitsburg, MD: U.S. Fire Administration. Retrieved from http://www.usfa.fema.gov/downloads/pdf/publications/tr_162.pdf
2. Stambaugh and Sensenig, *Special report: Fire department preparedness.*
3. DeSalle, R., ed. 1999. *Epidemic! The World of Infectious Diseases.* New York: New Press.
4. Ghosh, T. K. et al., eds. 2002. *Science and Technology of Terrorism and Counterterrorism.* New York: Marcel Dekker.
5. Tanaka, S. 1997. *The Buried City of Pompeii: What It Was Like When Vesuvius Exploded.* New York: Hyperion.
6. U.S. Environmental Protection Agency. 2007. *Incident Management Handbook: Incident command System (ICS).* Washington, DC: Author.
7. Infield, G. B. 1971. *Disaster at Bari.* New York: Macmillan.

24 Environmental Assessment

Pollution has occurred. Lingering pollution at contaminated sites is still a problem in many areas. River sediments once polluted by discharge of untreated wastes directly into streams and lakes still remain. Old dumpsites that were not constructed to isolate their waste still exist. Abandoned industrial sites still contain soils and buildings that were contaminated by previous spills, leaks, and improper handling of hazardous materials.

Many of the companies that polluted have disincorporated, and, as a result claim no responsibility for their past actions. Existing companies that are associated with contaminated sites (i.e., placed wastes in common dumping sites) and taxpayers result in being responsible for cleanups. Current environmental legislation (see Chapter 27) demands *strict and several liability* requires all responsible parties to pay for assessment and cleanup costs on the basis of ability to pay, not on the relative amount of waste that they individually "mismanaged." Laws are *retroactive*, punishing polluters for noncompliance of regulations even though such laws were not in effect when the pollution occurred.

Consequently, abandoned sites under *Superfund* laws are identified, *potential responsible parties (PRPs)* are identified, PRPs are officially notified of their legal responsibilities, and assessments and cleanups are contracted. The process is cumbersome at some sites, and progress can be slowed down with the involvement of attorneys. Nevertheless, environmental projects for cleaning up contaminated sites are taking place.

Environmental projects can involve tasks such as *site characterization*, *risk assessments*, *feasibility studies*, and *remediation*. For large and complex projects, tasks can be subdivided into subtasks. Research studies can help to guide the project to a successful conclusion.

Site Characterization

A site (piece of real estate property) might need to be assessed for environmental problems because it is known or suspected to be contaminated. Prior to a real estate purchase, the buyer or seller might need to determine if it is contaminated, even if no prior history of contamination is known or suspected. Ownership of land carries with it the owner's liability of environmental issues. If the seller uses due diligence prior to the sale to discover any environmental problems and divulges those problems to the potential buyer, the buyer can voluntarily assume those liabilities after the sale or chose not buy. A complete *site characterization*[1,2] might include:

A General Survey of the Site—Reconnaissance

Prior to a site visit, the investigator should gather information about the site from the published and unpublished literature, online information, and interviews. This might include books, journal articles, agency files, newspapers and magazines, and residents or employees who are familiar with the site or area. A brief visit to the site might be made, if safe, noting potential receptors (people in the area), and the pathways that contaminants might travel between the site and receptors.

Work and Other Written Plans—Plan Your Work in Order to Effectively Work Your Plan

A project-specific *work plan* should be written. The plan must specify exactly what tasks must be done in detail, with information on staffing, schedule, and budget. If sampling might be required, expected contaminants and types of *environmental media* (water, groundwater, soil, waste) should be specified (included or under separate cover).

Using information from the work and sampling plans, a health and safety plan must be written if there are potential or suspected hazards on the site. *Environmental assessments* usually involve health and safety issues. *Health and safety plans* might include the use of *personal protective equipment (PPE)* and clothing, setting up zones for *decontamination* of site personnel and the equipment used, health profiles of site staff, directions to hospitals, first aid supplies, etc. Also using information from the work and sampling plans, a quality assurance/quality control plan may be needed. Quality is assured if the tasks are monitored during all phases of the project.

FIGURE 24-1 Field measurement of surface water parameters

Noninvasive Testing

Sometimes aerial reconnaissance, geophysical, geochemical **(Figure 24-1)**, or remote sensing might be informative in gathering new site information prior to invasive testing leading to sampling. With *noninvasive testing*, observations can be made remotely or on the ground without disturbing the contaminated soil or surface waters. Therefore, exposure to contaminants by site workers is minimized.

Invasive Sampling and Testing

When additional information is warranted, representative sampling of the site soils, seeps, surface water, groundwater **(Figure 24-2)** and wastes can be obtained by digging test pits, drilling borings into the soil and rock, and installing groundwater monitoring wells.

The locations of samples must be documented by position surveying and accurate note taking. Samples must be taken according to *environmental agency protocols* with *quality control (Q/C)*. Quality control includes taking duplicates for independent analysis (two labs) and by submitting blanks that should either have no contamination, contain local uncontaminated media for background, or a fixed contamination (called spikes) to test laboratory *accuracy and precision*. *Blanks* may be used to determine if sample containers were contaminated between the site and the laboratory. Sample documentation should include detailed sampling procedures, date and time of sampling, shipment details including *chain-of-custody forms*, and time in transit to the lab before analysis—as some contaminants have a limited *shelf life*. The lab should know the expected contaminants, if known, and the relative concentration of the contaminants for equipment calibration and setup.

FIGURE 24-2 Groundwater testing

Analysis

A lag time may ensue prior to receipt of lab reports. When the results are received, they should be reviewed for accuracy, precision, and completeness. Precision differs from accuracy, as the former assesses the repeatability of the analysis from duplicate analyses. Accuracy, on the other hand, assesses the correctness of the results. If contaminants tested for are not detected, the lab results should indicate *N/D (nondetect)* with documentation of the *detection limit* of the instrument and official analytical method used. There can be no result marked "ZERO," as all analytical methods can detect down to certain levels, but hardly down to one molecule. Q/C problems, e.g., contamination during sample shipment, might cause rejection of the data or the sampling process must be repeated.

Report Writing

The site background, methods of study, results, and project conclusions of the site characterization must be communicated clearly to the intended party—the agency, client, seller or buyer of the site, etc. Data can be attached in appendices to the report. The report should be transmitted with a cover letter documenting its transmittal.

Risk Assessment

Depending on the outcome of the site characterization, a *risk assessment* can be performed to evaluate impacts to human health, ecological stressors, or dangers to the environment. The risk assessment includes the site characterization (from a study described in the section before), an *exposure assessment*, a *toxicological assessment*, and a *risk characterization*. First we must understand the concept of risk.

What Is Risk?

Risk is a condition involving the chance of occurrence.[3,4] It can be mathematically expressed as a *probability* from 0 (the condition will not occur) to 1 (the condition will definitely occur). In gambling, the risk is the probability of winning over loosing. A die has six sides, so each side has the probability of one sixth. A penny has two sides (heads or tails), so a head or a tail has the probability of one half. Betting at the roulette table on either red or black is actually less than one half, because two extra possibilities are zero and double zero.

FIGURE 24-3 Risk is a probability of occurrence

In dealing with dangerous exposures to contaminants one should manage the risks. Managing the risks helps to reduce or remove the danger or, in certain cases, accepts risks in exchange for a positive outcome. Consider the example of a health professional who enters a quarantined area where people are dying of a contagious disease, risking catching the disease herself, in order to administer medicine that would save the lives of others. In science, the endpoint risk values of 0 or 1 almost never occur. Fractional values between the endpoints ($0 < X < 1$) are common and express the

FIGURE 24-4 Will dangerous condition X happen?

likelihood of the outcome. The turtle must risk harm and extend its head and limbs (and rear parts) out of its protective shell to go about its life **(Figure 24-4)**.

Expressing Risk Quantitatively

If you gamble, you are taking a financial risk of losing your money. If you flip a coin repeatedly, the probability of it landing heads up is 0.5, a number halfway between 0 and 1. If the probability is 0.5, you might as well not gamble. All that work and worry and you just break even. All games of chance (where the casino takes a percentage off the top) have less than a 0.5 probability of winning, so the moral is simple: do not gamble at a casino.
Financial investments in businesses can give you better odds of gaining a profit (maybe >0.5), but there still is risk involved. In exploration, you risk your safety to find new lands, planets, and life forms. In education, you risk time and money to improve your skills and intellect. In research as well, you risk disappointments until you may find cures for diseases, new inventions, and explanations of fundamental observations.

Exposure to danger in the environment can cause harm to us, to other living creatures, and to the environment itself that sustains us. We must assess these risks, communicate the risk to others, and manage the risks to prevent or mitigate adverse consequences. If we do not take some risks, like the turtle, we cannot go about our lives.

Risk as a Probability of Occurrence

In environmental science, risk can be expressed in different ways. It can be a probability number from 0 to 1. Probability numbers are used in characterizing the risk of cancer incidence. Genetically, without specific reference to the risk of exposure to specific chemical contaminants or physical hazards, one out of three people (probability of $\frac{1}{3}$) is likely to develop cancer within 70 years of one's lifetime. In excess of that probability, exposure to carcinogens (agents that cause cancer) can produce an *excess cancer risk (ECR)*. The *US Environmental Protection Agency (USEPA)* has set targets for ECRs based on the receptor group that may be in harm's way. As a general rule, the ECR probability is 1/million, meaning that the danger of more than one extra person of the population getting cancer from one carcinogen in one pathway cannot be greater than $\frac{1}{3}$ + 1/1 million, or 0.330001. That seems like a very small extra risk from additional carcinogen exposure. If you were that one in a millionth person who gets cancer, however, you would not think of this target risk as small. You would probably suggest a 1/billion target risk for ECR. Then, the billionth person who gets cancer may have a different perspective . . . and so on.

Other Quantitative Expressions of Risk

Quantitatively environmental risk can also be expressed using other numeric parameters. USEPA research has determined that humans can be exposed to certain concentrations of contaminants without harm. They call each of these maximum values a *reference dose (RfD)*. If the environmental exposure concentration to these contaminants is greater than the RfD, then a threshold is reached that spells danger to the health of the receptor. If the exposure concentration is equal to or less than the RfD, then this does not represent a health hazard. Therefore, a new value is defined called the *hazard quotient (HQ)*, which is the value of exposure concentration divided by the RfD. If the HQ is greater than 1, there is a health hazard. RfD is used with contaminants in water. A reference concentration, RfC, is used similarly for contaminants found in air. The HQs are cumulative—that is, we can take the sum of all HQs and define a *hazard index (HI)*. The HI is the sum of all HQs from each contaminant and from the same contaminants reaching the receptor from all environmental media (surface water, groundwater, air, waste, soil, etc.).

We use another quantitative measure of risk in environmental science, the *lethal dose (LD-#)* value. This is important in distinguishing between toxic substances. The science of toxins (poisons) is called *toxicology*. The Renaissance physician known as Paracelsus (1493–1541) was a pioneer in this field. He amended his famous line that any substance could be a poison with the caveat that it depends on the dosage. A tiny amount of botulism toxin can kill many people in a short period of time. Dosages of lead can also be lethal, but one would need to administer a greater quantity of lead than botulism toxin. Sugar can be lethal in even greater dosages. Even pure water can be lethal in great quantity, as it can critically dilute electrolytes.

The *lethal dose (LD)* among substances would carry a number that is inversely proportional to its toxicity (virulence). The most toxic would have a small number and the least would have a large number. This is because the LD is the dose that would cause a certain percentage of a test population (or certain body weight) exposed to the substance to die. *LD-50* means a concentration that would kill 50% of the population tested. Lethal doses can also be expressed with different populations or body masses (LD-10, LD-1, LD-99, etc.) and with different exposure times and delays of treatments. Therefore, the lower the LD-50 (or LD-10, etc.) dosage, the stronger the poison.

Qualitative Expressions of Risk

We use *comparative risk assessment* when we cannot assign a number to the risk—when we cannot give it a probability of occurrence or compare it numerically to an RfD or LD number. One source of risk can be *greater or lesser* than another source of risk. Bicycle riding carries less risk than sky diving. Drinking milk carries less risk than drinking alcohol. Living near a forest preserve carries less risk than living near a nuclear power plant. Of course, you can think of risks for each of the lesser risk scenarios from these examples. Bicycle accidents can be serious. Lactose-intolerant people can be harmed by ingesting milk. Forests can be the source of deer ticks (lime disease vectors) and rabid raccoons. But these risks are still lower than those posed by the alternative scenarios.

Risk Perception (Not Objective)

Risk evaluation is not necessarily objective. Subjectivity enters into the evaluation when we perceive a greater or lesser risk than is scientifically valid. Our fears and prejudices influence our objectivity of *risk perception*. Here are two scenarios:

1. Statisticians tell us that flying in an airplane is less risky than driving in an automobile. Mile for mile, there are fewer deaths from plane crashes than from automobile accidents. However, when a jet crashes, mostly all passengers die (and lots of them at one time). During the "final descent" of the doomed flight, each passenger must feel helpless, especially when crammed into the economy section like sardines. Each passenger is powerless to help the cockpit crew in this time of crisis. In an automobile, however, the speeds are slower and the vehicle is on the ground, providing shorter stopping distances and a feeling that you, the driver, are in control of the vehicle and the situation.
2. A freight train carrying nuclear waste passes through your town. The risk of a derailment and dumping of radioactive material into the environment, the statisticians tell us, is very slight. Unfortunately, if the accident does occur, the severity of the outcome is catastrophic.

   Both of these examples—air travel and radioactive waste transport—indicate very low risk, but very high risk perception.

Risk Tolerance

To manage our risk, we must understand our *tolerance for risk*. This is based on our perception of risk. If this was the stock market in volatile economic times, we might not have the stomach for wild swings in stock market prices. If the *Dow Jones Index* has large, frequent drops, do we sell at the first sign of a recession and buy back at a boom time? That would be a mistake, as we would buy at high prices and sell at low prices, the opposite of what we should do. Consequently, risk perception can work to one's disadvantage. In studying risk assessment, you must keep an objective frame of mine.

Risk Arithmetic

As you can see, quantitative risk assessment requires some mathematics. Let us review some arithmetic.

If we use probability terms, it is important to recognize numbers between 0 and 1. This can be tricky if we use different mathematical notation. Yes, fractions are easy; a number between 0 and 1 is expressed by a fraction where the numerator (number above the fraction line) is smaller than the denominator (number below). For the fraction to be a probability, the fraction must also be positive.

The fraction

$$2/5$$

can be a probability. The fraction

$$6/5 \text{ or } -2/5,$$

is not. 6/5 is greater than 1, and $-2/5$ is a negative number.

In decimal form, a probability number, again, must be positive. Furthermore, the number to the left of decimal point must be 0. Number or numbers to the right of the decimal point can be any value between 0 and 9.

In environmental health we also deal with scientific and engineering notation. This involves exponents, numbers in the superscript position (upper right after the number). For example,

$$2^6$$

represents two to the sixth power, or $2 \times 2 \times 2 \times 2 \times 2 \times 2$.

$$17^3$$

represents seventeen to the third power, or $17 \times 17 \times 17$.

$$1.76 \times 10^5$$

represents 1.76 times $10 \times 10 \times 10 \times 10 \times 10$, or 176,000.

Remember, any number with a zero exponent equals 1. Also, any number with an exponent of 1 merely equals the number itself. So,

$$400^0 = 1,$$

and

$$400^1 = 400.$$

When the exponent is a negative number (important in environmental risk assessment), then that means that a fraction is formed with the numerator equal to the number 1 and the denominator equal to the same number having the exponent, but with a positive exponent. For example,

$$10^{-5} \text{ equals } 1/10^5.$$

In environmental risk assessment, we also do arithmetic with numbers having exponents. For example, if we add two numbers with the same power of 10, such as

$$(2 \times 10^3) + (3 \times 10^3)$$

we add the numbers before the powers of 10 and keep the same power of ten and get

$$5 \times 10^3.$$

What if we must add

$$(2 \times 10^3) + (3 \times 10^4)?$$

Three and four are different exponents. Yet,

$$3 \times 10^4 \text{ is the same as } 30 \times 10^3$$

$$(3 \times 10 \times 10 \times 10 \times 10 \text{ equals } 30 \times 10 \times 10 \times 10).$$

So, we add 30 and 3 and keep the power of 10^3, yielding the answer

$$33 \times 10^3.$$

We also can multiply and divide numbers with exponents. We must do this with the same numbers to the left of the exponent, usually powers of 10, and having the same or different exponents. For example, in

$$10^4 \times 10^5,$$

we add the exponents yielding

$$\mathbf{10^9}$$

Because 4 + 5 = 9.

To divide numbers with exponents, namely

$$\mathbf{10^8 \text{ divided by } 10^2},$$

we subtract the exponents yielding

$$\mathbf{10^6}$$

Because 8 − 2 = 6.

In engineering notation, we use the letter capital E instead of the 10 to express powers of 10. For example,

$$10^5 = E + 5;$$

$$10^{-5} = E - 5.$$

$$1.76 \times 10^{-8} = 1.76\ E - 8.$$

In this way, we avoid superscripts and multiplier ×'s. All arithmetic manipulations follow the same rules. Easy?

We will have more practice with real situations with risk numbers later in the chapter.

But before we move on, do you know which of the following are valid risk numbers (between 0 and 1)?

$$\mathbf{0.1, 1.1, 10^2, 10^{-2}, -0.75, 6^{-35}, 1.2\ E{-1}?}$$

Risk Assessment Rationale

A risk assessment is an evaluation of potential risks. An environmental health risk assessment evaluates the potential adverse health risks from contaminants and pathogens in the environment.

Why do we use risk assessments? We use them to predict the outcomes of the effects of exposure to contaminants and physical hazards. From these predictions, we can warn the public of the dangers from exposure, which constitutes risk communication. If we know the origin of the risk, we can remove or mitigate the source of the risk, block the pathways in the environment that are spreading the problem, or protect or remove the public from exposure. This is the aim of *risk management*. After risk management, we can look back at the original (*baseline*) risk assessment to evaluate the effectiveness of the cleanup process **(Figure 24-5)**.

FIGURE 24-5 Risk planning

Is There Always a Risk?

A contaminant or physical hazard may or may not produce harmful effects depending on:

1. Concentration of the contaminant or severity of the physical hazard.
   Low concentrations of toxic substances may not cause harm.
2. Toxicity of the contaminant (LD-50).
   Weakly toxic substances (high LD-50) may not cause harm.
3. Duration of the exposure.
   A short duration of exposure may not cause harm.
4. Frequency of exposure (small amounts from repeated exposures vs. large amounts all at once).
   The body can recover from a single exposure better than repeated "attacks" by the contaminant.
5. Body weight of the receptor.
   Large body size (other things being equal) can better counteract exposure.
6. Target organ for the contaminant.
   Although the whole body may be large, a specific target organ may be small. A large elephant has a small brain, smaller than a human's. As a result, a fixed exposure of contaminants targeting the brain would affect the elephant to a greater or equal degree than the human.
7. Carcinogenic vs. noncarcinogenic effects.
   All carcinogens can also have noncarcinogenic effects. Noncarcinogenic effects usually have thresholds, below which no adverse effects are observed. Carcinogens will have adverse effects over time even in very small doses (no threshold). There is some risk of cancer with any exposure.
8. Time it takes to develop adverse effects.
   Acute exposure (less than 2 weeks) is usually felt immediately if the dose threshold is reached. Chronic exposure (greater than 7 years) or subchronic exposure (2 weeks to 7 years) may take years to develop.
9. Synergistic factors from multiple contaminants.
   Sometimes adverse conditions resulting from the exposure to two contaminants do not equal the sum of the conditions resulting from each, added together. One plus one may not equal two. Sometimes the combined effect is greater or lesser.
10. Body's own defense to fight the hazard.
    The body can fight back. It can expel the contaminant. It can repair cell damage. It can kill a pathogen (e.g., bacterium, virus) before damage is significant. It can kill malignant cells that are spreading at the expense of benign cells.

How We Do a Risk Assessment[5]

There are four main *parts to* a risk assessment.

1. *Data collection and evaluation*
2. *Exposure assessment*
3. *Toxicity assessment*
4. *Risk characterization*

Data Collection and Evaluation

The risk assessment is only as good as the quality of the data collected. You have probably heard of the expression, "Garbage in, garbage out." A positive statement would read,

"Quality data in, quality assessment out." This is true about risk assessments. The USEPA maintains standards of testing contaminants from various environmental media. The agency specifies how samples are gathered, preserved, transported, analyzed, and validated. *Data quality objectives* are defined for each project, and quality assurance protocols are followed.

Great care and confidentiality are maintained with samples from contaminated sites. In addition, duplicate samples *(dupes)* are gathered for analysis and sent to different laboratories. Unknown to the laboratory, some samples are added to the shipment that are clean *(blanks)*. Clean samples are used to see if contamination was caused during the transport to the laboratory or even in the laboratory upon initial preparation. Other samples are "secretly" spiked with known chemicals (*spikes*) to test the sensitivity of the analytical methods.

When the results come back from the laboratory, duplicates, spikes, and blanks are looked at first to see if the laboratory data are reliable. Some of the samples will have the notation ND (nondetect) instead of a concentration number. ND means that contamination was not detected. There may be no trace of the contaminant in question, or the method used might not be able to detect such small quantities of the contaminant. It could also mean that contamination was barely detected, but a concentration number could not be assigned.

How do we use ND data in a risk assessment? ND could never mean that there definitely was zero concentration, as analytical procedures cannot detect down to one molecule of the substance. Sometimes we use one-half (or other fraction) of the detection limit, assuming that there was some of the contaminant present at an undetectable level.

After *data validation*, some data might be deemed unacceptable and thrown out. If more data are needed, additional sampling may be mandated.

When all validated data are assembled and deemed sufficient for the risk assessment, the second part of the assessment can begin.

Exposure Assessment

We are exposed to a contaminant by coming into contact with it. At the point of contact the contaminant can enter our bodies in several ways: *ingestion* (eating or drinking), *inhalation*, *absorption* through the skin surface, or *injection*. An exposure assessment evaluates how the receptor or receptors might be affected by the contaminant.

If a highly toxic substance exists and there are receptors to be exposed, but no pathway to deliver the contaminant for entry into the body, then there is no exposure. If there is exposure, then the risk will depend on various site conditions:

1. Physical setting of the site (e.g., climate stressors, soils)
2. Potential exposed populations (e.g., population size, healthy adults, children, elderly, infirm, full-time residents, occasional visitors)
3. Pathways to receptors (e.g., air, water, waste, vectors)
4. Point of exposure from the environment (e.g., faucet, sandbox, tailpipe, stream)
5. Changes in the concentration or toxicity of the contaminant in transport to the receptor
6. Frequency of exposure, or number of times contact is made
7. Duration of exposure, or length of time for each contact
   a. Chronic (7–70 years)
   b. Subchronic (2 weeks–7 years)
   c. Acute (<2 weeks)
8. Contact rate, or amount of soil, water, air, etc., contacted per unit time or event

9. *Magnitude* of contaminant, or amount of the agent available at the point of entry into the body
10. *Routes of entry* into the body (e.g., ingestion, dermal absorption, inhalation, injection)

From these data, a *chronic daily intake (CDI)* can be developed for each contaminant and for each environmental medium. CDI is determined by the equation

$$I = C \times [CR \times EFD]/BW \times 1/AT$$

- C: *chemical concentration* over exposure period (amount contaminant per amount of matrix)
- CR: *contact rate* (amount of contaminant media per unit time or event)
- EFD: *exposure frequency (EF)* and *duration* (ED) (days/year; days)
- BW: *body weight* (kg)
- AT: *averaging time* (days)

This equation can be modified depending on the type of exposure. The contact rate for eating contaminated fish would depend on how often the fish were eaten and how big a portion. Swimming in contaminated waters would depend on time in the water and body surface area. Air toxics exposure would depend on breathing rate during various activities.

Toxicity Assessment

The USEPA toxicological data may be used to identify hazards from numerous contaminants based on RfDs or RfCs and cancer factors for carcinogens. These data and background information on how these data were compiled may be found online (open and free to everyone) on the Integrated Risk Information System (IRIS). Data were compiled from animal testing and disease statistics on human populations (*epidemiological data*).

You do not have to be a toxicologist to understand the IRIS website, but a few definitions are helpful.

First, let us look at a *dose–response curve* (**Figure 24-6**). The horizontal axis shows arbitrary units of dosage. The vertical axis shows arbitrary adverse effect units. Points on the chart indicate pairs of dose–response data. Let us say you ingested an alcoholic beverage, an ale, pint by pint. Each dosage unit represents one pint. Depending on your body weight and physiology, no noticeable effects occurred until after the fifth pint. Then subsequent pints steeply increased the adverse effects. If the sixth pint showed an effect, but the fifth did not, your threshold value is between 5 and 6. We cannot determine from these data where between 5 and 6 the actual effect started.

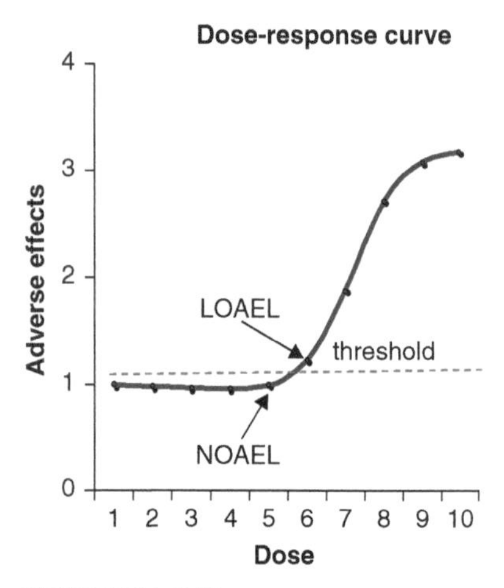

FIGURE 24-6 Dose–response curve (courtesy of Alan Jacobs)

The fifth dose then is defined as the *NOAEL* (pronounced *no ale*), which stands for *No Observed Adverse Effect Limit*. The sixth dose is defined as the *LOAEL* (*low ale*), which stands for *Lowest Observed Adverse Effect Limit*. This is the dose–response curve for noncarcinogenic effects.

When the NOAEL is known, it is used as the basis for the RfD or RfC rather than the higher threshold value and the yet higher LOAEL. If only the LOAEL is known, however, the RfD or RfC is made more conservative (lowering the LOAEL dose). The LOAEL is divided by an *uncertainty factor* in calculating the RfD or RfC.

The dose–response curve for carcinogenic effects is different (**Figure 24-7**). Here the horizontal axis is the dose and the vertical axis is the risk (probability value). Experimental values are for high doses, as the data are from animal testing. Animals used are laboratory mice that have a 2-year life span. To develop tumors in the mice in this short span, high doses must be administered. Obviously, this is not the scenario for human development of cancer, where low doses over decades of time may be required for tumors from carcinogens to develop. To extrapolate to the human scenario, the curve is extended to the left at lower risk levels and then further modified to be a straight line ending at zero risk. This results in uncertainty for the results when animal data are extrapolated to humans.

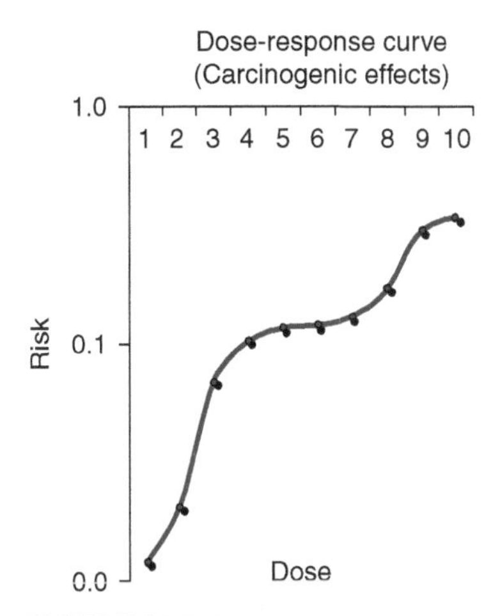

FIGURE 24-7 Dose–response curve for carcinogenic effect (courtesy of Alan Jacobs)

An additional piece of data is developed for carcinogenic effects called the *Weight of Evidence* classification. In a risk assessment, one must list the class (A through E) according to the evidence known:

A. Human carcinogen (human data)
B. Probable human carcinogen
   B1: limited human data
   B2: sufficient animal data, inadequate or no evidence in humans
C. Possible human carcinogen
D. Cannot prove human carcinogenicity
E. Evidence for noncarcinogenicity for humans

Risk Characterization

The fourth and final part of a risk assessment is the risk characterization. This part combines the first three parts (identification, exposure, and toxicity) and determines whether there is actually a risk to human health. A similar process can be performed for health impacts on nonhumans (wildlife) or the environment.

A. Review outputs from the exposure assessment.
B. Calculate the CDIs separately for noncarcinogenic and carcinogenic effects.
C. Review outputs from the toxicity assessment by compiling the RfDs and RfCs and the cancer slope factors for carcinogens.
D. List the weights of evidence for all carcinogens.
E. Calculate the hazard quotient for each contaminant in each environmental medium (CDI/RfD or RfC). Then add up all the HQs to calculate the HI.
F. Calculate the excess cancer risk for each carcinogenic contaminant in each environmental medium (CDI × slope factor). Then add up all the excess cancer risks.
G. A risk characterization cannot be considered complete unless the numerical expressions of risk are accompanied by explanatory text interpreting and qualifying the results (with uncertainty factors and weights of evidence).

Testing the Results of Risk Assessments

Chronic exposures involve small doses over long periods of time. To test the results of a risk assessment, we need to revisit the exposure scenario at a later time to see if the results are borne out. With respect to predictions of cancer incidence, we need at least 20 years after the initial exposure and longer, as tumors would develop at least a few decades after the initial exposure. Also, because many diseases are caused by factors in addition to the contaminants used in the risk assessment, one needs to compare populations exposed to these contaminants with those who were not exposed.

Superfund Aftermath

The USEPA started identifying abandoned sites that were contaminated by various chemicals after the agency's formation in 1970. Thousands of contaminated, abandoned sites have been considered and documented, and the worst of these sites have been prioritized on the *National Priorities List (NPL)*. Some sites have been *remediated* (cleaned up) and others have not. Since the sites were contaminated, releases of hazardous materials have found environmental pathways (surface and groundwater, seeps, transported soils and sediments, air, waste) to the receptor's point of exposure. From risk assessment predictions performed on these sites, a certain percentage of the population should have developed diseases from their exposure. Have they?

The *Centers for Disease Control and Prevention (CDC)*, the *National Cancer Institute (NCI)*, the *National Institutes of Health (NIH)*, and state and local health departments have compiled information on disease incidence, county by county, and state by state, in all 50 states. From toxicology and epidemiology studies, we know about many of the contaminants and the diseases they cause. Specifically, we know about the parts of the body that can exhibit adverse health effects from each contaminant. For example, lead can attack the bones and nervous system, benzene can cause leukemia, and mercury compounds can cause birth defects.

Is there a greater incidence of certain diseases in areas near sites that are contaminated with certain chemicals known to cause those diseases[6]? If we look at the incidence of liver-bile cancer in counties of Maryland and Delaware, we see an interesting correlation. One compound called vinyl chloride is a known *human carcinogen* that targets the liver. Counties where vinyl chloride had been released into the environment 30 years ago have a greater incidence of liver-bile cancer than counties where this contaminant is not known. These sites happen to be NPL sites that have vinyl chloride as a contaminant of concern. This situation is also found in other states (e.g., Massachusetts, Washington, New Jersey). Not all states, however, show this correlation. Hawaii, for example, has the highest incidence of liver-bile cancer in the United States, although there are no known vinyl chloride–contaminated sites in Hawaii. Why do some states show correlations, whereas others do not? There are several reasons. One is because there is another carcinogen that targets the liver, namely, arsenic, which is found naturally in the soil in various concentrations throughout the United States. Another is because other diseases of the liver, due to diabetes, alcoholism, and an overabundance of iron, can also lead to liver-bile cancer. Finally, our genetic makeup can be the cause of a greater propensity to contract certain cancers.

Although there are many causes of liver cancer (as well as causes of other diseases) as a result of exposure to chemicals from contaminated sites, a lesson can be learned from risk assessments. We may be naturally susceptible to certain diseases, but exposure to specific chemicals can increase our chances of contracting those diseases. There is a greater risk with increased exposure. The environmental cry of "Not in my backyard!" (*NIMBY*) rings true for these substances.

Humans are not solely responsible for environmental risks. Nature contributes sources of pollution, disease, radiation, toxins in soil and rock, and catastrophic events that are life-threatening.[7] We excuse Mother Nature's sources. Why?

Left to its own laws, Nature can solve environmental problems as they arise. Succession can reestablish ecosystems. Natural attenuation of contaminants can remediate accumulated wastes. Decomposer organisms can recycle vital macro- and micronutrients. Floods replenish soil nutrients. This is the concept of *Gaia*, where Earth can heal itself. Lake Erie, once considered dead, recovered to again become a healthy ecosystem.

Because Nature plays no favorites, we must be on guard for the sources of environmental risk for which Nature is responsible. For example, we can encounter poison ivy (skin rash), venomous snakes, bees, and deer ticks (carriers of Lyme disease) in the forest. We can be exposed to West Nile virus and malaria protists from mosquito bites. Bites from infected dogs, cats, bats, and raccoons can give us rabies. We can be exposed to harmful UV radiation from the sun. We can contract lung cancer from radon gas emanating from cracks in the Earth. Our environment can be polluted from gases and particulates ejected from volcanoes. We cannot eliminate many of these sources, but we can take proper precautions to guard against their adverse effects on our health and safety.

Added to these natural sources of environmental risks are the risks created by human actions. These sources are called *anthropomorphic* (literally human formed). Most of these sources are derived from the mismanagement of hazardous substances and wastes.

References

1. HMTRI. 1997. *Site Characterization—Sampling and Analysis.* New York: Van Nostrand Reinhold.
2. J. Ravi, L.V. Urban, G.S. Stacey, and H. Balbach. 2002. *Environmental Assessment*, 2nd ed. New York: McGraw-Hill.
3. Kammen, D. M. and D. M. Hassenzahl. 1999. *Should We Risk It? Exploring Environmental, Health, and Technological Problem Solving.* Princeton, NJ: Princeton University Press.
4. Rodricks, J. V. 2007. *Calculated Risks: The Toxicity and Human Health Risks of Chemicals in Our Environment.* New York: Cambridge University Press.
5. U.S. Environmental Protection Agency. *Risk Assessment Guidance for Superfund (RAGS), Part A.* Retrieved from http://www.epa.gov/oswer/riskassessment/ragsa/
6. Jacobs, A. M. and S. Waldinger. 2011. *Vinyl chloride contamination in aquifers at U.S. NPL sites correlated with incidence of liver cancer* (Abstract). Fourth International Conference on Medical Geology, Bari, Italy, September 20–25, 2011.
7. European Geophysical Society. 2001. *Natural Hazards and Earth System Sciences*, Vol. 1, no. 1/2. Katienburg-Lindau, Germany.

25 Remediation

Image © Rob Wilson, 2014. Used under license from Shutterstock, Inc.

Results of an environmental assessment are used to determine the best course of action to clean up the contaminated sites.[1,2] The first step in the remediation process is to do a *feasibility study*.

Feasibility Study

A *feasibility study* can follow an environmental assessment.[1] This study proposes alternatives for dealing with the problem. Included in all feasibility studies are two alternatives that do not require a remediation (cleanup). They are (1) *no action*, and (2) keeping receptors away from the contaminated site by *administrative controls* (legal action). Obviously, these alternatives do not remove the problem. No action produces no risk reduction. Administrative controls might not be permanent and future risks might arise.

These alternatives (do nothing, or do little) are used to compare costs and effectiveness of the other alternatives that would be proposed, that would offer greater risk reductions.

Other *alternatives* involve some kind of remediation, e.g., isolating the contamination source, physically blocking the pathway away from the source, and/or protecting potential receptors from exposure to the environmental risks. Other alternatives can use established strategies that must reduce three things: (1) the volume of contaminants or contaminated media, (2) the toxicity or other hazards, and (3) mobility (ability for the contaminants to spread).

Examples of remedial alternatives are: capping the site to prevent leachate generation, building impermeable barriers to hem in contamination, incineration of wastes, removal of contaminated soil and waste and place it in a properly engineered landfill, pump out contaminated groundwater and treat it—putting clean water back into an aquifer, treating contaminated aquifers underground by injecting air, pH-buffering liquids, or nourishment for contaminant destroying bacteria.

For the specific site, feasible alternatives are compared to each other with respect to cost, ability to be implemented, effectiveness, speed of the cleanup, prevention of adverse side effects, and compliance with existing laws. Problems could arise from the implementation of certain alternatives; for example, (1) if volatile organic compounds could be removed from groundwater and emitted to the air, increases in air pollution may not be acceptable; (2) if mixed waste is incinerated, more toxic compounds (like dioxins) might result and unburnable wastes (metallic compounds) would require burial in a chemical waste landfill; or (3) landfilling of wastes might produce methane that is a greenhouse gas.

After a careful comparison of all alternatives, the best one is proposed. If accepted, a cleanup can be begun.

Site Cleanup

Implementation of a cleanup may take a long time. During the cleanup, strict health and safety rules must be maintained to prevent harm to site workers performing remediation tasks (e.g., drilling, earth moving, pumping of contaminated groundwater, digging test pits) and other potential receptors in the vicinity of the site. *Public relations* may be important, as the public in the cleanup area would be concerned seeing activity, perhaps by people in "moon suits."

Follow-up tasks might include a postremediation risk assessment to calculate *risk reduction*. Risk reduction is a good measure of the success of the remediation. Also, *periodic monitoring* may be performed (e.g., groundwater sampling and analysis) to make sure that the effectiveness of the cleanup is long-term.

References

1. O'Brien & Gere Engineers, Inc. and Robert Bellandi. 1988. *Hazardous Waste Site Remediation*. New York: Van Nostrand Reinhold.
2. Grasso, Domenic. 1993. *Hazardous Waste Site Remediation – Source Control*. Boca Raton, FL: Lewis.

UNIT 7 Sustainable Environmental Solutions

26 Waste Management

We must not repeat the mistakes of the past by continuing to increase and miss-handle our wastes.[1,2] Each person in the United States produces about 5 pounds of solid waste and 70 gallons of liquid waste each day. With a growing population the total amount of waste grows as well. To avoid additional sources of waste generation, we should practice *waste minimization* or, as a goal, complete *pollution prevention (P2)*.

Waste Minimization

Recycle

Recycling minimizes the amount of waste produced.

One should recycle containers and objects that have reached their useful life. Such materials that are metallic, glass, plastic, paper, or cardboard may be used as raw materials for new product manufacturing. Recycling reduces our need for additional mining of metals, production of glass from sand, refining petroleum for plastics, and destroying trees for paper products.

Many communities in the United States have recycling programs. Recyclable materials are deposited in *drop-off bins* by residents or taken to the *curbside* for pick-up in front of their residences. Collected recyclables are then transported to recycling sorting facilities and then sold to companies that process them for new products. Recycling in the United States is voluntary, but in European Union countries it is mandatory, with fines levied for noncompliance. Land for landfills in Europe is scarce and expensive, so trash is inspected and valued (called *valorization*) for potential recycling **(Figure 26-1)**.

FIGURE 26-1 Colored-glass recycling bins in Leipzig, Germany. The color of the bin matches the color of the glass

Other materials that can be recycled include garden waste and food waste. This can be accomplished by a process called *composting*. These wastes are placed in a bin or a pile outdoors for several months. Periodically, the pile is turned over to allow air and moisture to flow through the mass. Sometimes mineral soils are added (e.g., sand or clay). It is inadvisable to compost meat and fats, as vermin and anaerobic bacteria are attracted to the pile that could result in disease and bad smells. One can, however, perform *worm compost* on meats and fats indoors, requiring manual decanting of the worm wastes periodically. Recycling of food and garden waste is sometimes done in the United States, but in Europe this is routine and also mandatory.

Household hazardous wastes, however, are collected by municipalities for treatment and disposal and not recycled.

In industry, recycling also has been practiced by companies. *Scrap metal* has been a valuable resource in manufacturing. This reduces disposal costs and the purchasing costs of some additional raw materials. Potential wastes are recycled by the same company in their processes or sold to other companies who need those by-products for their processes. The manufacturer can also minimize the waste outputs by adjusting their processing methods. Recycling turns waste into a resource.

Reuse

Recycling requires sorting and possibly cleaning before the material can be remanufactured into other products. This requires human work and energy to clean, sort, and remold the material into a new product.

A better way to minimize waste, if possible, is to *reuse* products. If the article or container can be cleaned up and reused, remanufacturing is not necessary. A glass jar of cooked fruit can be washed and used to store food or nonfood articles. Plastic or paper bags can sometimes be reused.

In the early twentieth century, milk was delivered to residences in glass bottles and empties were collected, washed, and refilled with milk. Nowadays, milk is sold in paper or plastic cartons that, if not recycled, end up in landfills. Beverages in glass bottles were bought in groceries with a deposit charged on each bottle, with empties brought back to the grocery for the deposit refund; returned empties were washed and refilled. There are some states that charge a deposit as an incentive for the customer to return the beverage bottles—but nowadays the glass bottles (thinner these days) are recycled by melting down the glass and remanufacturing new bottles.

Shopping bags and shipping containers can be made for reuse. So-called *box stores* in the United States and mostly all groceries in Europe do not provide free bags. You can buy plastic bags (some biodegradable) individually or bring your own reusable bags for transporting your purchases home. Some Americans accept the free bags at groceries and then reuse them as garbage or storage bags. In Europe and in some US shops, customers buy and bring with them cloth shopping bags that are reused during the next shopping trip, saving plastic and paper.

Manufacturers could save money by reducing the amount of packaging or provide reusable containers that the customer could return to the place it was purchased for a reward.

Unfortunately, nonbiodegradable plastic bags that are indiscriminately discarded and end up in the sea can cause aquatic animals to get tangled in them. These bags are also a suffocation hazard.

Reuse has been around for a long time. Within a family, hand-me-downs were common. Materials can be bought for re-use at: junk yards, pawn shops, garage/tag sales, used furniture (maybe, antique) shops, flea markets, secondhand clothing stores, clothing donation organizations, consignment shops, rummage sales, used-book shops, silent auctions, and estate (contents) sales. Used or reconditioned products, restored dwellings, and used cars provide markets for consumers at lower prices.

Reduce

It is easier said (written) than done, but reduction in the amount of goods we buy can also *reduce* the amount of waste we produce, including products and packaging. If we really need a product, then, all other things being equal, choose the product with the best chance of it being recycled or reused and having the least amount of packaging.

Pollution Prevention

Pollution prevention is the ultimate goal. If you do not produce waste, you do not need to clean it up or even recycle it. It is not there. The easiest way to do this is to continually increase recycling, reusing, and reducing our generation of waste. What waste is left should be converted into a resource. Mother Nature is the only existing system on Earth that has achieved this ideal.

Waste Treatment and Disposal

Waste sources include those discussed in Unit 5 (Environmental Problems). Waste can also be generated during remediation of a contaminated site, as discussed in Unit 6 (Correcting Environmental Mistakes). If wastes pile up, we must treat and/or dispose of the waste.

Both Mother Nature and our technology can help.[2]

EPA regulations that cover waste generation dictate rules for the temporary storage, treatment, and disposal of wastes. Wastes that are hazardous have more stringent regulations. Waste generators may need to send their nonhazardous solid wastes to a licensed *municipal solid waste (MSW) landfill* or *incinerator* and their liquid wastes to a wastewater treatment facility (a *publicly owned treatment works—POTW*). If their liquid wastes are hazardous, there may be some on-site pretreatment required. They might send their hazardous wastes to a licensed *treatment, storage, and disposal facility (TSDF)*. What happens next depends on the nature and quantity of the waste materials.

Natural Attenuation

Some wastes are handled by Mother Nature because there are always bacteria or fungi around capable of metabolizing some constituents and making them less hazardous to us and to the environment. Also, natural Earth systems can filter, dilute, disperse, and chemically alter waste to make it less toxic, less voluminous, and less mobile. When nature processes and reduces the waste, we called that *natural attenuation*. It is usually a slow process and not necessarily 100% effective. Human intervention can be a good thing to speed things along.

One example of human intervention to a natural process is at so-called *land treatment units*. Hazardous waste is applied directly on the soil surface or into upper soil horizons. Sunlight or microbes can act upon the contaminants to degrade them into nonhazardous materials. These units are not required to have liner systems or leachate collection and removal systems. Sometimes, dozers churn up the soil and waste (similar to composting) to allow a more complete degradation to occur. Naturally, in land treatment units, the site and equipment used must be monitored to be free of contaminants after the human-assisted, natural attenuation is completed.

This system was successful in treating diesel oil contamination of riverbank sediments from a storage tank spill (800,000 gals) on the shores of the Monongahela River, south of Pittsburgh, Pennsylvania. The spill also affected water and sediment in the river and downstream reaches of the river system.

Waste Treatment

We can mimic and augment natural processes to expedite the treatment of waste. Numerous technologies have been developed. Some treatments reduce the waste characteristics to make them less or nonhazardous so the residual products from the treatment can be disposed of more cheaply and with less long-range liability.

Treatment technologies include those briefly mentioned in the discussion of feasibility studies in proposing remedial alternatives for a cleanup of contaminated sites. Many of these technologies have adverse side effects, so safe disposal or additional treatments may also be required. For example, incineration, which usually burns up more than 99% of the mostly organic waste, can create air pollution problems from stack emissions. Incineration also generates unburnable ash that usually has hazardous characteristics.

Air or steam stripping of volatile organic compounds from contaminated groundwater is another technology. Groundwater that is pumped to the surface for treatment is processed in specially constructed towers filled with gravel—and steam is added to some. The process vaporizes the volatile organic compounds in the water and frees them to escape at the top of the towers into the atmosphere. Then, the cleaned water is pumped back into the underground aquifer. An adverse side effect might be an unacceptable increase in the concentration of these volatiles in the local atmosphere.

The *stripping process* is only one technology that requires pumping groundwater to the surface for treatment. So-called *pump-and-treat technologies* can have the pumped water treated by reaction with ozone, radiated with ultraviolet light, or filtered by activated charcoal.

Another technology is the in situ (in place) treatment of groundwater in subsurface aquifers, eliminating the pumping of groundwater to the surface. In this process, air (*air sparging*) or bacterial nutrients are pumped into the aquifer (one type of *bioremediation*) to initiate oxidation or bacterial digestion, respectively, of organic contaminants in the aquifer.

Contaminated water is not the only medium that can undergo a cleanup. Soil can be washed to separate out undesirable metals that have a greater density from less dense nonhazardous silica sand or silt. Certain species of plants, which uptake toxic metals from the soil are planted. They are removed afterwards for burial, leaving the soil in a cleaner condition—a process called *phytoremediation*. Contaminated air can be filtered, cleaned by electrostatic precipitation, or dispersed into the atmosphere where permissible.

A POTW can handle most wastewater that comes from our toilets and drains. This is discussed in detail in Unit 3 (Sustainable Material Resources). POTWs can also handle some of the waste (pretreated sometimes) from industrial waste generators.

Discharge of wastewater (mostly after treatment) into natural surface waters, needs to have approval from environmental regulatory agencies. Regulations that cover such discharges must be permitted under the National Pollutant Discharge Elimination System (NPDES permits). At the point of discharge, the water is not necessarily fit for consumption. The local situation must allow for additional, natural attenuation and pretreatment (e.g., chlorination) by public water distribution companies.

Disposal

We cannot match the thoroughness of Mother Nature in waste treatment. After all our efforts are made to minimize and prevent waste, there still exists waste that must be isolated. Isolation of solids is done by burial; isolation of liquids is done by solidification followed by burial, or by deep well injection. Incineration is also a disposal alternative.

Municipal Solid Waste (MSW) Disposal

Prior to the development of regulations, garbage (or ashes from local incinerators) was dumped in mounds at the edge of town, in the ocean (by near-shore waste generators), or privately buried on vacant land. All this activity was done without concern about the mobility of or public harm from the waste or waste constituents after dumping. Consequently, dump sites allowed waste to sit below the water table, contained waste placed on permeable soils over aquifers, or uncovered to spread wind-blown debris away from the site. Old tires and undrained depressions were sites of mosquito breeding. Humans traversing the site (for scavenging) were subject to injury from sharp objects and contagion from pathogens (especially from medical wastes). Hazardous waste was mixed in with nonhazardous waste. Liquid waste was mixed with solid waste.

We now regulate the disposal of waste. *Municipal solid waste* (MSW) is that which is picked up at curbside or in dumpsters and sent to a licensed MSW landfill. Most of these landfills are found in counties not far from the population they serve. The MSW landfill is engineered to keep the solid waste intact, and liquids generated from the solid waste (*leachate*), if any, from seeping below the base of the landfill. Leachate, which is

FIGURE 26-2 Municipal solid waste

generated when rainwater percolates through the solid waste and carries contaminants deeper in the landfill, is prevented from leaking out the bottom by the placement of plastic and clay liners. Leachate is also pumped out periodically and treated or recycled back into the landfill to allow further decomposition of the solid waste. The waste is covered with clean soil daily. Groundwater is monitored with wells, especially down the water-table gradient from the site.

When the landfill reaches its volumetric limit, it is capped with clay and vents are installed to prevent buildup of waste-decomposition gases. Some landfills let the vented methane (a greenhouse gas) escape and some few collect the methane as a fuel.

Hazardous waste or liquids are not allowed in MSW landfills. As all the wastes from curbsides or dumpsters are not inspected, some disallowed wastes get in.

There are separate landfills for *construction/demolition wastes.*

Food and beverage containers, packaging, used paper, etc. if put in landfills is a waste of space **(Figure 26-2)**. If these same materials are recycled or reused they can be a useful resource. A manufacturer can recycle or sell some of the by-products of its processes rather than pay disposal costs. A municipality can incinerate the organic material (with proper air-quality precautions) in its solid wastes, using this waste as a fuel to generate electricity and use left-over heat to warm adjacent offices. Methane gas, generated from organic waste by anaerobic bacteria, can be used as a fuel. Used cooking oils from restaurants can be used as a vehicle fuel (with proper engine adjustments).

A true waste is material that cannot be used again. Nature does not recognize this definition, as all natural wastes are recycled or reused. We can learn from Nature . . . we must sustain our limited resources.

Hazardous-Waste Disposal

There are special sites designed to accommodate the disposal of hazardous wastes.

These sites include specially designed landfills, *surface impoundments* (temporary), waste piles (temporary), *injection wells*, and *underground repositories*. Surface impoundments and injection wells are the main disposal sites for liquid wastes. No liquids are allowed in landfills, so *solidification* (encased in concrete or mixed with cement) or *vitrification* (fused into glass) is necessary before landfill burial.

Hazardous-waste Landfills

Landfills that can legally accept hazardous wastes are similar in concept to MSW landfills. The big difference is that they require more safeguards to prevent contaminants from escaping to the environment. These safeguards include double liners, double leachate collection and removal systems, *leak detection*, drainage and wind dispersal controls, and intense monitoring, quality assurance, and inspection. Consequently, these landfills number less than two dozen in the United States and have much higher disposal fees. Shipping waste to these sites that are farther way from the sources of waste also require high transportation costs, with the shipping requiring more safeguards (e.g., placarding, chain-of-custody forms, leak-proof transports).

Surface Impoundments

Temporary storage (or holding for treatment) impoundments that can safely hold liquid waste are of various kinds (e.g., natural depressions, pits, ponds, lagoons). They must have safeguards similarly found at hazardous-waste landfills.

Waste Piles

Piles of solid waste are noncontainerized and temporary. They must have safeguards similarly found at surface impoundments. Drainage and wind dispersal controls are required similar to those at hazardous-waste landfills.

Injection Wells

Disposal of liquids down injection wells requires a well system that is deeper than existing sources of groundwater that may impact springs, surface drainage, mines, and groundwater aquifers. Such impact might pollute water supplies. Injection wells that are very deep may also present environmental problems. Earthquakes can be triggered at deep crustal faults. Earthquakes have been recorded from deep-well injection of chemical wastes at the Rocky Mountain Arsenal in Colorado. Also, recent low-Magnitude earthquakes have been linked to injection of wastes from shale-gas exploitation using hydrofracturing.

Radioactive-Waste Repositories

Radioactive waste from electric-power generating stations and the military is a special category of hazardous waste. The radioactivity cannot be treated, is highly hazardous, and will be highly hazardous for thousands of years. The strategy for this type of waste is to isolate it in deep rock formations in caverns mined out to contain it. The sites were named *repositories*, with the hope that the radioactive material could be retrieved if it would be needed as a resource in the future. In fact, these so-called repositories are permanent burial sites, so the word "repository" is a misnomer.

For radioactive military waste, a site has been established in the Carlsbad, New Mexico, area in deep caverns mined in a 2,000-foot-thick salt bed, 2,150 feet below the ground.[3] Called the *Waste Isolation Pilot Project (WIPP)*, it is the only nuclear-waste repository in the United States and one of only three in the world. At the beginning of 2014, low-level leaks of radioactive air were detected and the project was temporarily (?) shut down.

The senior author visited the site underground after construction but before radioactive waste was placed in the caverns. Prior to the visit, preliminary evaluation of the salt bed predicted that water of hydration was low and cracks in the salt would be minor.

Our survey in the caverns and tunnels using a geophysical tool called *ground-probing radar (GPR)* indicated that there were numerous cracks in the salt. Other studies indicated that the walls of the mined-out caverns were starting to close in prematurely. After filling the caverns, the salt bed is supposed to eventually close in around the waste (salt creep) as a permanent seal—this was happening sooner than expected.

Aside from military radioactive waste, commercial nuclear waste is piling up at each of our nuclear power plants, with no repository planned or constructed as a permanent disposal site. Waste includes spent fuel pellets, fuel rods, water from the reactor pool, and reactor-site equipment.

It was hoped that deep caverns mined out of volcanic rock in Yucca Mountain in Nevada[4] could be constructed so that commercial nuclear waste could start to be buried there. The caverns are deep, yet above the water table, in a geographic area that has low rainfall. Volcanic activity had occurred more than 5 million years ago. The public along the routes from the current storage sites (about 120 of them), however, are apprehensive of the transport of nuclear waste through their communities (Not in my backyard—NIMBY) on route to Yucca Mountain. Environmental organizations were also fearful of the site because of seismicity in the area (earthquake potential).

The relative risks at the current storage sites (at the power plants) should be weighed against the risks at Yucca Mountain (or another site), and we should quickly establish a central repository as soon as possible. The current storage sites could be targets for terrorism and human-caused accidents. Many of those other sites are in the vicinity of many populated areas, and some have earthquake potential similar to that at Yucca Mountain.

Unfortunately, the Yucca Mountain site or an alternate site will not be established soon. Recent troubles at the WIPP site and the public's perceived fear of radioactivity do not bode well for reactivation of the project. Additional funding to complete and start-up the Yucca Mountain facility was halted in 2010.

References

1. Bishop, P. L. 2000. *Pollution Prevention: Fundamentals and Practice.* New York: McGraw-Hill.
2. Wentz, C. A. 1995. *Hazardous Waste Management*, 2nd ed. New York: McGraw-Hill.
3. National Research Council. 1996. *The Waste Isolation Pilot Plant: A Potential Solution for the Disposal of Transuranic Waste.* Washington, DC: The National Academies Press.
4. Walker, J. S. 2009. *The Road to Yucca Mountain: The Development of Radioactive Waste Policy in the United States.* Berkeley, CA: University of California Press.

27 Environmental Regulations and Social Responsibility

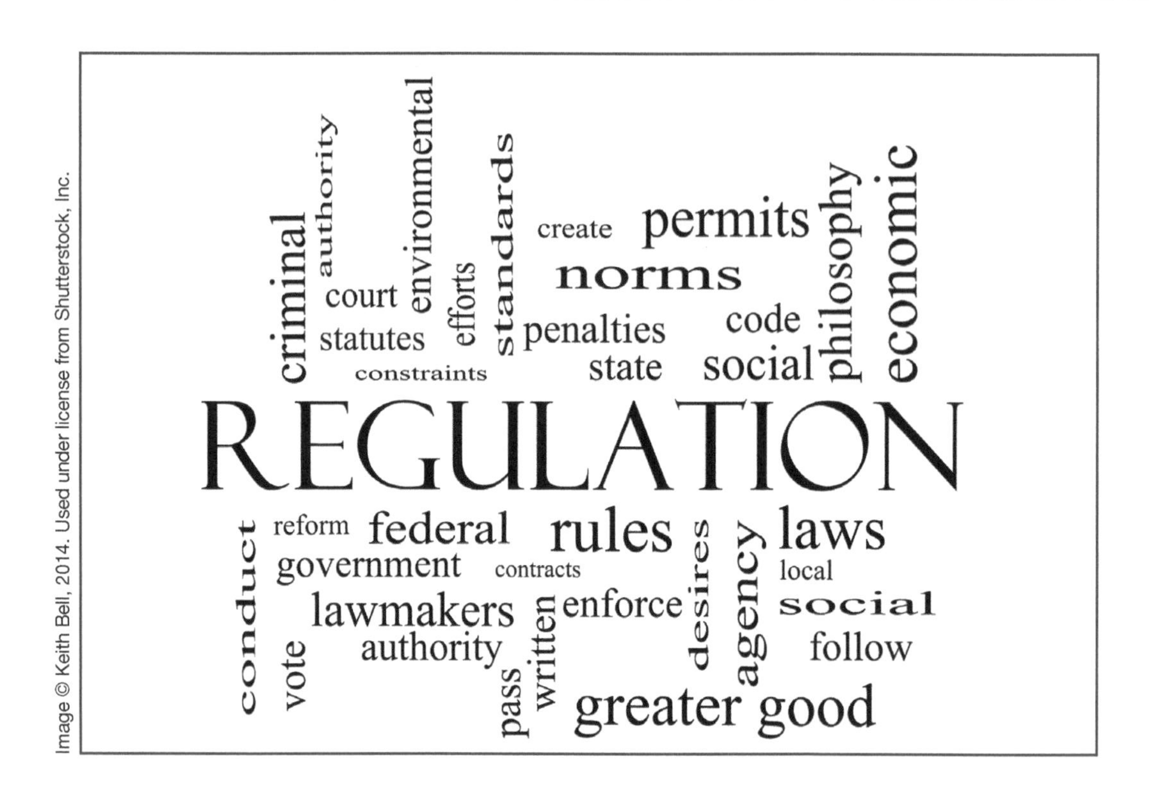

The Origin of Effective Regulations

Prior to the establishment of the federal and state environmental protection agencies (before 1970), society had noticed but, in many cases, tolerated environmental degradation. By the 1960s, however, there was a growing awareness that we were poisoning our environment. Finally we are linking these poisons in the air, water, and soil to resulting adverse health and safety effects. The Cuyahoga River in Ohio caught fire due to floating flammable oils. Air in all major cities was laden with smog. Indiscriminate dumping of wastes was common, and health problems were reported around chemical dump sites like Love Canal, New York, and Times Beach, Missouri. Mayor John Lindsay of New York City from 1966 to 1973 jokingly responded to reporters' questions concerning the poor air quality of the city to the effect that: "I never trusted air that I could not see."

Environmental degradation became more obvious and serious because of the growing population (then exceeding 3 billion people worldwide), advances in manufacturing, consumerism, technological advances in chemical engineering (e.g., plastics, fertilizers and pesticides, pharmaceuticals), and an absence of regulations in the management of wastes. Note that the population reached 7 billion people in October of 2011.

Awareness was helped by the publication of several popular books, including *Silent Spring* (1962) by Rachel Carson and *The Population Bomb* (1968) by Paul Ehrlich. Carson alerted her readers to the adverse health effects of pesticides and Ehrlich emphasized the problems of overpopulation. These books jarred our consciences into questioning our pursuit of wealth at the expense of health.

Prior to 1970, locally adopted ordinances had few federal laws that supported local efforts to clean up the environment. Nevertheless, there were some federal laws (*acts of Congress*) prior to this date that began to address environmental issues. These included the following[1]:

FIFRA (1947) Insecticide, Fungicide, and Rodenticide Act
CWA (1948) Clean Water Act
CAA (1955) Clean Air Act
CAA (1963) Clean Air Act, Title 2 (for Motor Vehicles)
SWDA (1965) Solid Waste Disposal Act
AQA (1967) Air Quality Act
NEPA (1969) National Environmental Policy Act (for federal projects).

Unfortunately, the environment was still being degraded. There was still a need for amendments to the existing laws and new laws, detailed regulations, and enforcement of the rules with stiffer penalties for noncompliance. Furthermore, adverse health effects from contaminants in the environment were not well understood. Compliance was fair at best. Fines paid by companies that polluted the air, waterways and water supplies, and land (surface and subsurface) were low enough to be considered production costs. Without admitting guilt, the waste generators paid the fines and passed the costs on to consumers. Many companies refused to admit that their indiscriminate disposal or release of chemicals caused adverse health effects. Medical services for treatment of adverse health effects were borne by individuals, either directly or through their health insurance premiums.

Residential areas in or adjacent to industrial and waste disposal areas were subjected to health risks from the discharge of pollutants. These areas were mostly where low-income citizens lived. This resulted in health disparities between affluent and poor neighborhoods. Environmental health problems in poor neighborhoods included down-wind exposure to toxic air pollutants from industrial stacks, groundwater pollution from indiscriminate dumping of wastes, and inhalation and ingestion of leaded-paint chips and dust.

Legislation was needed to promote strict zoning and siting regulations, empower local citizens to have a voice in restricting the siting of certain industries in their neighborhoods, and reduce and control pollution in general. Citizens of low-economic status also had less access to affordable health care from environmental (and other) causes. These health disparities resulted in racial tensions and disenfranchisement of residents in neighborhoods that were exposed to a greater amount of adverse health effects from nearby sources of pollution.

Although pre-1970 laws and regulations did not stem the tide of environmental degradation, they were a start. In the coming decades, issues of *environmental regulation* and *environmental justice* were to be addressed.

Acts of Congress that are important to the environment[2] include several pertaining to the major environmental media. The act of Congress that addresses the problem of abandoned hazardous waste sites is the *Comprehensive Environmental Response, Compensation, and Liability Act (CERCLA)*. CERCLA was passed in 1980, with additional amendments, referred to as *SARA*, in 1986. This act established a fund, known as the Superfund, for covering costs of sites where responsible parties could not be found. Regulations for CERCLA are in CFR 40 (Rules for the Protection of the Environment). Where possible, the USEPA aims to recover costs from parties that are responsible for producing the hazardous conditions at these sites. When sites are identified, they are prioritized and placed on the National Priorities List (*NPL*) according to the severity of the hazards for priority attention for cleanup.

Federal regulations are published in the *Code of Federal Regulations (CFRs)* and are organized into titles according to the agencies involved.[3] For example, safety rules for those working at hazardous sites are administered under OSHA in the Department of Labor (Title 29). Rules governing transportation of hazardous substances are administered by the Department of Transportation (Title 49). New chemicals that may be hazardous are regulated by the Department of Commerce (Title 15). Superfund site regulations supervised by the USEPA are published in the CFR under Title 40. State laws and regulations are published by individual states. The federal requirement on state laws and regulations is that they be as strict, or stricter, than federal laws and regulations.

Also under Title 40 are rules for the management of wastes from existing waste generators. The *Resource Conservation and Recovery Act (RCRA)*, passed in 1976, covers wastes from the time they are produced to the time they are disposed of. Metaphorically, this concept is known as "cradle to grave."

The act pertaining to newly manufactured hazardous substances is the *Toxic Substance Control Act (TSCA)*, with the acronym pronounced like the opera *Tosca*. TSCA was enacted in 1976. Regulations are published in CFR Title 15 (Department of Commerce).

The Safe Drinking Water Act (*SDWA*), establishing public drinking water standards, was enacted in 1974. Regulations define primary (health protective) and secondary (addressing bad tastes, odors, and discolorations) standards. These regulations apply to supplies that serve at least 25 people for at least 60 days per year and have 15 or more service taps.

Chemical contaminants in the water are measured in parts per million (ppm). Permissible limits have maximum concentrations (*MCLs*) for each chemical. The MCLs are not completely health protective, as frequent analysis and removal for some contaminants present in concentrations of fractions of a part per billion (one thousandth of a ppm) would be cost prohibitive by local water suppliers. Local supplies are required to test for sewage contamination (coliform bacteria presence) daily.

The Clean Air Act (*CAA*), as amended most recently in 2012, was a revision of the former *Air Pollution Control Act* of 1955 and subsequent amendments. It addresses air pollution prevention and control, emission standards for moving sources, acid deposition control, permits, and stratospheric ozone protection.

The United States was a pioneer in environmental protection, and many other developed countries used United States waste management standards as a model for their own laws. Some countries, especially those with greater population densities (e.g., those in Western Europe), were anxious to develop waste management standards because they have a scarcity of land to use for landfills.

As environmental problems can extend beyond national borders, the global environmental problems were tackled by convening international conferences. These included the following[4]:

1987 Montreal Protocol (limiting chlorofluorocarbon production)
1992 Earth Summit in Rio de Janeiro

1994 U.N. Conference on Population in Cairo, Egypt
1997 Kyoto, Japan, Accord (limiting greenhouse gas emissions)

While the United States regulations are under the purview of the federal government, states can oversee studies and cleanups of contaminated sites and environmental media (water, air, and soil). Many NPL sites in qualified, designated states that have a recognized program of regulation and enforcement can take the responsibility of overseeing the cleanups.

Brownfield revitalization is another area where states can assume primary responsibility. Companies are encouraged to clean up abandoned industrial sites (brown fields) with acceptable risk-based cleanup levels suited to subsequent intended use of the site.

Concerning water quality standards, the USEPA has jurisdiction over "*public water systems (PWS)*." A PWS, as defined by the USEPA, is " . . . a system for the provision to the public of water for human consumption through pipes or other constructed conveyances [that] has at least fifteen service connections or regularly serves at least twenty-five individuals." A further caveat is added "for 60 days a year" to eliminate transient water supplies. Public drinking water systems are regulated by the USEPA, and delegated states and tribes. The USEPA approves a state when it demonstrates compliance with USEPA codes and exhibits the ability to enforce its laws. The state may further delegate authority to local health departments for some programs, e.g., to regulate private residential water systems, for example.

Left completely to the states are regulations regarding the disposal of medical or infectious wastes. Such wastes are generated by medical services, testing, and research (humans and animals). They could include blood, tissue, and equipment or supplies contaminated with such biological materials. Prior to leaving the medical facility, such wastes may be incinerated or autoclaved prior to offsite disposal.

How Small Is A Part Per Million (ppm)?

A part per million (ppm) is a very small amount. In water, it means a drop of contaminant for each one million drops of uncontaminated water. A part per billion (ppb) is 1,000 times smaller than a ppm.

To fathom this minute amount, think of liquid drops as sports fans in a football stadium. A drop of a contaminant is a fan of the opposing team. A drop of uncontaminated water is a fan of the home team. If the stadium holds 100,000 fans and the team plays 10 home games per year, the stadium, filled to capacity each game, would house 1 million fans. If one of the fans in only one of the games in that season rooted for the opposing team, that would be one part per million (ppm) "contamination." Now if the stadium was the venue for 1,000 seasons and was filled to capacity each season (for 1,000 years), and only one fan rooting for the opposing team showed up for only one game in 1,000 years (10,000 games), then that would be one ppb of "contamination." Analytical methods can detect some contaminants in minute quantities down to parts per trillion or smaller. Such analyses, one would expect, would be very expensive.

Command and Control

In the decade after the establishment of environmental protection agencies, antipollution laws (enacted by legislative bodies) and regulations (rules developed by agencies for abiding by the laws) increased in number and strictness covering many aspects of

FIGURE 27-1 Regulatory signs

environmental protection. (**Figure 27-1**). Current waste generators were required to comply with new laws and regulations. Waste-dump sites were identified and the identities of the polluters were pursued. At the time of this writing (2014) there were approximately 1,300 sites in the United States that needed to be cleaned up.

The Woodlawn Landfill, in northeast Maryland, is an example of one of the numerous Superfund sites.[5] This 40-acre tract was identified by the U.S. Environmental Protection Agency (USEPA) in 1987 and placed on the National Priority List (NPL). Sites on this list have top priority for remediation. The Woodlawn Landfill was placed on the NPL because wastes from the site were polluting the groundwater. One of the wastes, sludge, was found to contain a carcinogen called vinyl chloride, known to cause liver cancer. The sludge was derived from wastes generated during plastics (PVC) manufacturing. At the time when the sludge was dumped, the carcinogenicity of vinyl chloride was not understood. Ironically, the waste generator received a permit from the State of Maryland to dispose of the *PVC sludge* onto a site that did not have protective leachate containment. The site was a former gravel pit that was used later as a municipal garbage dump. The sludge was placed in the dump. The resulting leachate, containing the vinyl chloride, infiltrated the permeable ground surface and seeped into the drinking water aquifer below and to surrounding, private water-supply wells. Today, such disposal must be located in engineered landfills containing impermeable liners and leachate collection systems, continuously monitored for leakage.

After 1970, laws and regulations for Superfund sites demanded that the parties involved with the disposal take responsibility for site cleanup, even though no laws or regulations were violated at the time of dumping (retroactive liability). The state permit for the dumping at the Woodlawn Landfill was not a valid excuse. The new regulations were retroactive.

Furthermore, laws and regulations now demanded that all parties (except the state) contribute to the cleanup without regard to their share of the blame for dumping. This "deep pockets" approach was valid because the new laws demanded "joint and several liability." Even if one of the parties had only a 1% involvement, they might end up paying 100% of the costs if they were able to pay and others were not.

These laws and regulations were strict enough to discourage waste generators from disposing wastes indiscriminately. The USEPA through the *Department of Justice (DOJ)* was now able to levy significant fines and even prison terms to persistent polluters. When a company would be sentenced to a prison term, a designated staff member was incarcerated. This employee was not necessarily the company president or CEO, but a mutually agreed upon person to serve on the company's behalf.

This heavy-handed justice procedure is called *command and control*. It was not totally effective because large corporations had access to legal consultants, who could slow down the process of enforcement of the cleanup through the judicial appeals process.

Not all laws pertain to waste management; other laws pertain to hazardous materials prior to their becoming wastes. Not all laws are at the federal level; states, counties, and municipalities also regulate hazards in the environment through their departments and agencies. Not all federal regulations are the responsibility of the USEPA; other federal agencies regulating environmental concerns include the Nuclear Regulatory Commission

(NRC) which oversees nuclear power plants and radioactive waste and materials, and the Departments of the Interior, Agriculture, Energy, Defense, Transportation, Labor, and others, according to their area of responsibility.

With the establishment of the USEPA, states have developed their own agencies for the protection of the environment. These agencies have a variety of names, such as the Ohio Environmental Protection Agency, the Pennsylvania Department of Environmental Protection, the Arizona Department of Environmental Quality.

Additional federal legislation from 1970 forward included the following:

CAA—Clean Air Act (1970) Amendments
OSHA (1970) Occupational, Safety, and Health Act
CWA—Clean Water Act (1972)—revised
CAA (1973–74) more amendments
SDWA (1974) Safe Drinking Water Act
HMTA (1975) DOT regulations for transport of hazardous materials
TSCA (1976) Toxic Substances Control Act (new chemicals)
RCRA (1976) Resource Conservation and Recovery Act (for current waste generators)
CAA (1977) more amendments
Solid and Chemical Waste Acts (1970–80)
EPCRA (1986) Emergency Planning and Community Right-to-Know
PPA (1990) Pollution Prevention Act
OPA (1990) Oil Pollution Act
SBLRBRA (2002) Small Business Liability Relief, Brownfield Revitalization Act

Significant events throughout the period of environmental awareness and beyond caused by human error include the following:

1969—Cuyahoga River fire
1978—Love Canal evacuated
1979—Three Mile Island nuclear accident
1984—Bhopal (India) pesticide leak disaster
1986—Chernobyl (Ukraine) nuclear accident (**Figure 27-2**)
1989—Exxon Valdez oil spill
1991—Persian Gulf War oil dump

FIGURE 27-2 Radiation danger sign at Chernobyl (Ukraine) accident site

2010—Deepwater Horizon oil spill
2011—Fukushima (Japan) nuclear accident (human error involved siting the plant at both plate tectonic boundary and tsunami danger zone; and ineffective backup cooling system)

Not included here are events that were not caused by human error, but have had environmental consequences. These include major earthquakes, tsunamis, landslides, volcanic eruptions, floods, droughts, and pandemics.

Market Forces

A more effective approach to waste management involves economics. Working with industry, the regulatory agencies showed the companies that waste minimization and pollution prevention (P2) were good for business. Industrial processes were modified to reduce and recycle wastes within the manufacturing facilities.[6] Also, wastes might be sold as resources to other companies. Such reducing and recycling was extended from the manufactured product to the packaging of the product. These practices reduced manufacturing and disposal costs.

When corporations became known for their respect for the environment (green companies), this reputation was publicized to their customers and shareholders. When environmental standards were established, companies could advertise their green status.

Environmental In-Justice

Perhaps better referred to as *environmental unfairness*, this situation tends to expose the poorer segment of society to more pollution than the general population.[7] Factories, mills, and landfills tend to be located in or adjacent to the poorer neighborhoods (once labeled *the other side of the tracks*). The poor have little political clout in controlling the zoning of land to geographically distance themselves from potential polluting industries. They also find it hard to move to other less-polluted inner-city areas because of higher housing costs there. If they move to rural or suburban areas, the poor may experience higher housing costs and commuting expenses. If health problems arise from exposure to pollution, health-care costs can be unaffordable to the poor.

Lifestyle

There exist health clubs that have decent parking lots adjacent to their facilities. Yet, some offer valet parking. People did not want to walk from their cars to a place where they would exercise on machines or do laps in the pool or around the indoor track. If patrons found their own parking places, they wanted those close to the entrance.

In buildings with elevators, people would wait for an elevator to go down one floor, with the staircase adjacent to the elevator.

Can we improve the environment for ourselves and for others? For our own environment, it may require a change in our lifestyle. This is very difficult for many. It might mean:

a. More carpooling
b. Switching to public transportation
c. Walking or biking to work
d. Reduction in noningested water use

e. Taking elevators only to go up more than one floor, or down more than two floors.
f. Switching to less toxic cleaning or gardening chemicals
g. Abstinence from smoking
h. Healthier diet
i. Less paper use
j. Fewer throw-away containers
k. Turning off lights and appliances when not in use
l. Lowering the thermostat in winter, and raising it in summer
m. Buying from "green" companies
n. Reusing grocery bags
o. Drinking fewer sweetened beverages
p. Drinking less alcohol
q. More recycling, reuse, and reduction of consumer goods
r. Composting
s. Enforcement of strict zoning laws
t. Promotion of environmental fairness
u. Avoidance of big lawns; adopt a landscape that is in harmony with your local biome
v. Switching to organic produce
w. Switching to buying locally grown products (less energy use for shipping)
x. Installing solar panels on roofs
y. Installing cisterns for nondrinking water supply.

Costs of Sustaining our Planet

Let us say that we "run up a tab," charges for all goods and services we use, itemized for food, clothing, furnished living space (rent or mortgage), utilities (electric, gas, water, and sewer), transportation (personal and public), recreation, education, medical and dental, trash collection, and other services and items. All of these items have an environmental impact either directly (charged to consumer) or indirectly (costs passed on to consumers), or have hidden costs (damage to the environment and health costs from pollution).

Out of our income, we spend money for products and services. In a totally free economy, when products and services are scarce and the demand is great, the prices go up. When items are plentiful and the demand is low, the prices should go down—but do they?

No economy is totally free (of control). Sometimes production is withheld by cartels, our taxes pay for subsidies to suppress production, tariffs are charged to importers, or prices are fixed by government action to prevent prices from going down. The consumer pays the bill.

Indirect Charges

No business is sustainable if the company loses money or does not generate profits for the company investors; otherwise shareholders could invest their money elsewhere. Nonprofits should not amass investment income, but they still must meet their expenses. Even though the consumer is not directly charged for the costs of production, the consumer pays these charges contained in the prices he or she pays. For example, the company must pay their electric bill, real estate taxes, waste disposal, etc., so those costs are part of the price passed on to the consumer.[8]

Hidden Costs

If there were no environmental regulations, companies could ignore costs for proper waste disposal and have a market advantage over companies that pay to properly dispose of their wastes. Environmental regulations promote a level playing field when requiring companies to mitigate and prevent pollution in the environment.

Illegal dumping and uncontrolled discharge of wastes to the environment, however, result in costs to the general public in the form of tax increases to pay for cleanups and medical (and medical insurance) costs from adverse health effects from the pollution. We assume that the air we breathe is free, but the clean air that we expect has hidden costs that someone must pay for to keep it clean. If the "free" air we breathe is not clean, we end up paying for the treatment of respiratory and other diseases.

What Is Our Footprint?

The term *footprint or ecological footprint* is the area on Earth needed to replenish our renewable resources and absorb the wastes generated.[8] An adult human's footprint should be larger than a child's. What about comparing human adults worldwide?

It turns out that adults of different countries and economic status have very differently-sized footprints. In the industrialized world, the "haves" have much larger footprints than the developing world "have-nots."

A photograph of an exhibit in the Swedish Universeum natural history museum in Gothenburg illustrates these comparisons **(Figure 27-3)** from four countries: Have-nots include Columbia, South America, and Bangladesh, Asia. The haves include the United States and Sweden. The haves show footprints that are climbing up the wall (too big). The green footprints in the middle both with a soled shoe and a bare foot indicate what should be our goal for footprint size to maintain a sustainable planet (1.9 hectares or approx. 30,000 sq. ft.). That still sounds like a lot.

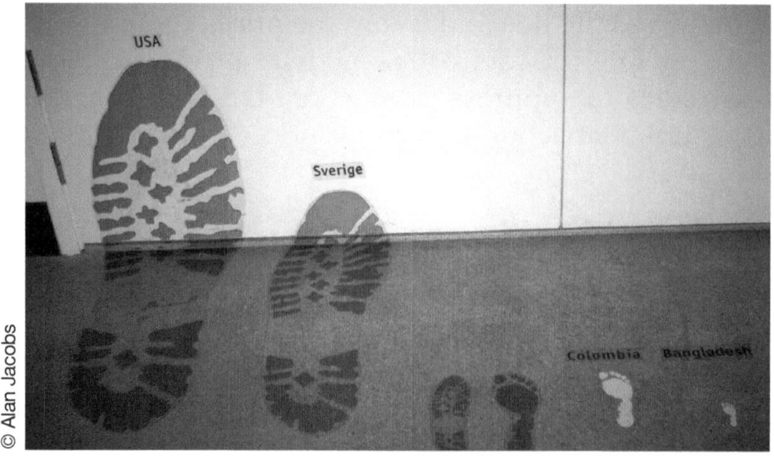

© Alan Jacobs

FIGURE 27-3 Ecological footprints: United States much larger than other countries

References

1. Jacobs, A. M. and D. Porter. 2009. "Environmental Health in Public Health." Chap. 24. *In Principles of Public Health Practice,* 3rd ed., edited by Scutchfield, F. D. and C. W. Keck. Clifton Park, NY: Delmar Cengage Learning.
2. Bregman, J. I. and R. D. Edell. 2002. *Environmental Compliance Handbook,* 2nd ed. Boca Raton, FL: Lewis.
3. U.S. Government, Code of Federal Regulations (CFR), Titles 15, 29, 40, 49.
4. DeSombre, E. R. 2000. *Domestic Sources of International Environmental Policy: Industry, Environmentalists, and U.S. Power.* Cambridge, MA: The MIT Press.
5. Jacobs, Alan M., I. E. Amin, and O. N. Fisher. 2007. Persistence of Vinyl Chloride in Ground Water at the Woodlawn Landfill Superfund Site, Northeastern Maryland, USA. *Environmental Geology* 52 (2007): 1253–60.
6. Bishop, P. L. 2000. *Pollution prevention: Fundamentals and practice.* New York: McGraw-Hill.
7. Walker, G. P. 2012. *Environmental justice: Concepts, Evidence, and Politics.* New York: Routledge, 2012.
8. Hardisty, P. E. 2010. *Environmental and Economic Sustainability.* Boca Raton: CRC Press.

Index